A FARM ON EVERY CORNER

A FARM ON EVERY CORNER

REIMAGINING AMERICA'S FOOD SYSTEM FOR THE TWENTY-FIRST CENTURY

DAVID LANGE

NEW DEGREE PRESS

A FARM ON EVERY CORNER

Reimagining America's Food System for the Twenty-First Century

ISBN   978-1-63676-534-1 *Paperback*

           978-1-63676-079-7 *Kindle Ebook*

           978-1-63676-080-3 *Ebook*

To Mom, Dad, Schlübbo, Emily, Molly,
Evelyne, Livia, Annie, and Adam.

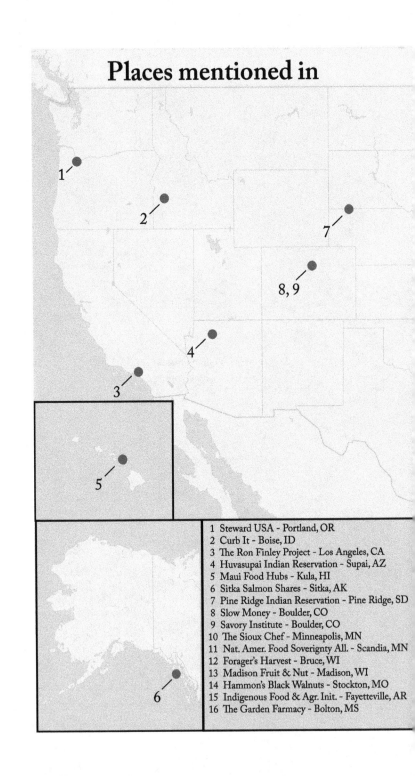

Places mentioned in

1 Steward USA - Portland, OR
2 Curb It - Boise, ID
3 The Ron Finley Project - Los Angeles, CA
4 Huvasupai Indian Reservation - Supai, AZ
5 Maui Food Hubs - Kula, HI
6 Sitka Salmon Shares - Sitka, AK
7 Pine Ridge Indian Reservation - Pine Ridge, SD
8 Slow Money - Boulder, CO
9 Savory Institute - Boulder, CO
10 The Sioux Chef - Minneapolis, MN
11 Nat. Amer. Food Soverignty All. - Scandia, MN
12 Forager's Harvest - Bruce, WI
13 Madison Fruit & Nut - Madison, WI
14 Hammon's Black Walnuts - Stockton, MO
15 Indigenous Food & Agr. Init. - Fayetteville, AR
16 The Garden Farmacy - Bolton, MS

A Farm on Every Corner

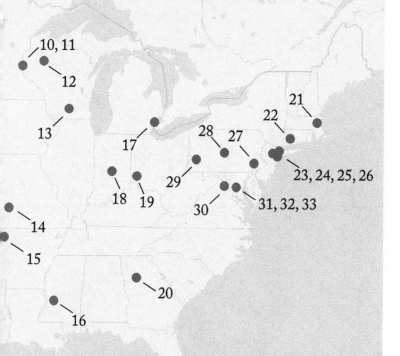

17 Fisheye Farms - Detroit, MI
18 Keep Indianapolis Beautiful - Indianapolis, IN
19 80 Acres Farms - Hamilton, OH
20 Browns Mill Food Forest - Atlanta, GA
21 Boston Natural Areas Network - Boston, MA
22 Ideal Fish - Waterbury, CT
23 Misfit Foods - Brooklyn, NY
24 Ducks Eatery - Manhattan, NY
25 Hunts Point Cooperative Market - The Bronx, NY
26 AeroFarms - Newark, NJ
27 Montgomery County Anti-Hunger Network - Pottstown, PA
28 Center for Economic & Community Development - State College, PA
29 Fruit Tree Planting Found. - Pittsburgh, PA
30 Ayershire Farm - Upperville, VA
31 US Department of Agriculture - Washington, DC
32 Seylou Bakery - Washington, DC
33 Sweetgreen - Washington, DC

CONTENTS

"We are each other's harvest;
we are each other's business;
we are each other's magnitude and bond."

—GWENDOLYN BROOKS

INTRODUCTION

A SONG OF SPROUTS AND STEEL

Walking down Rome Street on the ragged edge of Newark, New Jersey's Ironbound District, you would be forgiven to think you were in the wrong part of town. Overgrown plants reach out toward the broken-up, dusty sidewalk. A graffiti-covered highway overpass stretches over the pock-marked pavement, leading to a rusted truss bridge creaking under the weight of a passing train.

Rounding the corner, a group of boxy, nondescript build-ings materialize—an old steel mill long past its halcyon days. But, this emblem of post-industrial America has found new life. Past the newly-painted fence and through the renovat-ed concrete halls lies something equal parts anachronistic and stunning: the largest indoor vertical farm in the world.

Extending your gaze skyward, towers of trays filled with bright-green leaves inundated by LED light overtake your vision. As the building hums with the sounds of pneumat-ic pumps dispensing nutrient-rich vapor and workers on scissor lifts tending to the crops with care, you'd again be forgiven to think you were in the wrong part of the *country*. But the fact is this farm is producing vegetables that are cheaper and fresher than anything you can find in your average supermarket—and they're doing it in the middle of the inner city. In a place like this, the intersections of urban revitalization, cutting-edge technology, and the future of food collide to create nothing short of a miracle.

It got me thinking—if this marvel of agriculture and science can exist here, sandwiched between Route 9 and a community pool, it could exist anywhere.

But is this sustainable? Is this really a major shift in our economy or just a passing trend fueled by venture capital and "locavore" foodies? Can we really have a farm in every city? In every neighborhood? On every corner? Can we transform the way we eat by reconnecting it with where we live? Getting to the bottom of these questions on how to fundamentally change our relationship with food launched me on a quest. In essence, we can create a new, hyperlocal food ecosystem in every town across America to give us a more robust, resilient, sustainable, equitable, profitable, healthy, and tasty future.

A Growing Concern

Unlike many other post-industrialized nations, the United States is growing. By 2050, we can expect to have *over 438 million* Americans—an increase of over one hundred million from today.[1] This will be compounded by the savage

---

1   "U.S. Population Projections: 2005-2050," Pew Research Center, last modified February 11, 2008.

effects of climate change. The Intergovernmental Panel on Climate Change reports we can expect increased flooding in coastal and riverine communities, more and stronger severe weather events such as tornadoes and hurricanes, greater volatility in rainfall and droughts, higher temperatures and shorter winters, and more.[2] In the face of these challenges, the way we grow, consume, and dispose of our food needs to change.

Many people may think this means we need even larger corporate farms with stronger pesticides and more potent fertilizers. They may think we need to produce our food as cheaply and efficiently as possible or risk mass starvation. Who has time for organic on a warming, crowding planet?

The truth is these issues require innovative solutions and a rethinking of every facet of our food system. If we want to have a shot at coming out on top of these seemingly insurmountable problems, we need to think bigger by thinking smaller. Not smaller in the myopic sense, but rather creating new ways of doing things that work closer to home and work smarter instead of harder. Instead of filling thousands more acres with the same handful of crops (much of which will be either wasted in our fields, processing plants, and landfills; used as feed for animals; or turned into biofuel), why not pursue an abundance of fresh, local produce from the small farm that's

---

2   T.F. Stocker, et al. *Climate Change 2013: The Physical Science Basis. Contribution of Working Group I to the Fifth Assessment Report of the Intergovernmental Panel on Climate Change* (Cambridge, United Kingdom, IPCC, 2013).

just a short drive or walk away? Why not think local, or better yet, *hyper*local to find the answers to the pressing food problems of our time?

Global Problems, Local Solutions

When I got a job managing my local farmers market for the season, I was admittedly pretty unacquainted with the world of food. I had been to a farmers market before, but never thought much about where my food came from beyond "a farm out in California, probably." I was a recent college graduate without a job or much in the way of direction, and the gig seemed like a good deal—I'd only have to go into work Saturday mornings, after all!

Those six months of the market season I spent in that middle school parking lot turned out to be one of the most enlightening experiences I have ever had. Working alongside the people who grow and sell the food I eat every day and meeting my fellow neighbors who rallied around the market every weekend with a sense of hometown pride was illuminating.

But the most special thing to witness was the precious feeling of community that permeated every market day. Whether it was random folks helping their older neighbors to load up and carry their tote bags home or watching our pastry vendor sneak a little extra treat to the children of his regular patrons with a smile and a wink, it was apparent there was something extraordinary happening in this parking lot.

The idea food could become so entwined with the community when we localize our food systems became something of an obsession to me. So I devoured all the media I could find on it. The ramifications of what I was discovering was startling: how we make and from where we get our food had significant effects beyond just filling your plate.

Food touched on health, economics, politics, the environment, and our children, and often, in direct and tangible ways I could never have imagined. I came to realize the choices our governments, businesses, and households make every day in how we legislate, source, and consume food are beyond powerful: Employed efficiently and judiciously, we can uplift vulnerable communities, stimulate local economies, and work in harmony with our environment. From then on, I knew finding ways to use food to improve our health, our neighborhoods, and our planet was going to be a lifelong passion.

Change Begins Now

My journey to learn about the future of local food has opened my eyes to a whole new world of possibilities. All across America, amazing people are pioneering remarkable things in their communities. I believe exploring the innovators of today is crucial to discovering the scalable solutions that will make tomorrow possible. Applying what we've already learned and the incredible solutions the next few decades will yield to our food system can radically transform our relationship to food and to the systems that produce, distribute, and dispose of it.

From a food forest in Atlanta, to sidewalks in Los Angeles, to the docks of Sitka, Alaska, this book hopes to take you on a journey of what's possible when food meets community. We're going to take inventory of the current state of our nation's food system and make the case for why empowering our communities to take back food production from corporations makes us all better off.

So whether you're someone who attends their weekly farmers market like it's Sunday Mass or if you've never given a second thought to how your food got to your plate, this book is for you. Together, we can discover the potential located down the street and across the country and learn to reconnect how we eat with where we live.

PART ONE

INTRODUCTION TO THE FOOD SYSTEM

CHAPTER ONE

A BRIEF HISTORY OF AMERICAN AGRICULTURE AND FOOD SOVEREIGNTY

When it comes to our current globalized corporate food system, to understand how we got here, we need to know how things used to be done. I don't think it'll come as a surprise to anyone our food economy was not always like this (yes, there was a time before Whole Foods!) and it was not all that far in the past.

In a nation of over 2.2 million farms, just 125,000 United States farms produce 75 percent of all value in the agricultural sector. Many of these farms saturate their soil with fertilizer and plant thousands of acres with patented GMO seeds, destroying the environment from which the land came and turning it into nothing more than a vehicle for economic gain.

So how did we get here?

America's Agricultural Beginnings

Anyone who spent their formative years in the United States (and many who didn't) are familiar with the classic story of the founding and expansion of the United States: from thirteen colonies wrestling themselves from their British overlords to a new nation gallivanting Westward in search of land, wealth, and power.

But this common tale of glory belies the extremely gruesome reality of settler colonialism and its relationship not only to indigenous people, but the land as well.

American Indians, an extremely diverse term representing hundreds of tribal groups and millions of people who were the caretakers of the lands in what is now the United States, were slaughtered and exiled to reservations. This was to make room for development by homesteaders, capitalists, the military, and others who claimed the new *terra nullius* as their own. The American Trappist monk and famed writer Thomas Merton scathingly writes of European displacement of the original inhabitants of the Americas:

"They seemed to be owners of the whole continent, until we arrived and informed them of the true situation. They were squatters on land which God had assigned to us...We could see at a glance, we understood without the slightest hesita-

tion, that they were only aboriginal owners. They never had any legal title to the real estate."[3]

Rather than the aboriginal view of humans as stewards of their environment, the American mindset toward its land was not one of care toward a homeland, but rather a relationship built on exploitation and violence. The earth was not to be honored but plundered. Wendell Berry, considered to be one of the seminal writers of modern environmentalism, commented further on the difference between indigenous and colonizer relationships to the land:

*"The Indians did, of course, experience movements of population, but in general their relation to place was based upon old usage and association, upon inherited memory, tradition, veneration. The land was their homeland. The first and greatest American revolution, which has never been superseded, was the coming of people who did **not** look upon the land as a homeland... One cannot help but see the similarity between this foreign colonialism and the domestic colonialism that, by policy, converts productive farm, forest, and grazing lands into strip mines. Now, as then, we see the abstract values of an industrial economy preying upon the native productivity of land and people."*[4]

The bottom line always trumped any sort of compassion or sense of belonging toward the land—no surprise since American agriculture evolved alongside modern capital-

---

3   Matthew Fox, *A Way To God: Thomas Merton's Creation Spirituality Journey* (Novato: New World Library, 2016), 196-198.

4   Wendell Berry, *The Unsettling of America; Culture & Agriculture.* (San Francisco: Sierra Club Books, 1977).

ism and the industrial revolution. Traditional foodways of American Indian tribes who worked within local ecosystems to create bountiful, mixed harvests (see Chapter ten "Reclaiming Seeds and Knowledge in Indian Country") were steamrolled to make way for huge fields of non-native crops that extracted as much nutrients from the soil as possible until it was no longer useful. Industry came to the rescue, introducing more and more potent nitrogen and phosphorous fertilizers, diesel farming machinery, and eventually, designer seeds. Innovation after innovation was designed to pillage the land for all the productivity it could offer. Thus, to frame the American relationship with the land in the context of colonialism is to recognize this extreme philosophy of land usage has its roots in the very heart of American agriculture and capitalism.

The Evolution of the Modern Food System

Bearing this framework in mind, we can now turn to the past one hundred years in the evolution of the modern corporate food system. As settlers spread across the American continent, they built their own local food systems along the way. This was typically in the pursuit of some high-value trading good, such as furs in the Northeastern woodlands, bison in the Midwest, cotton in the Southeast, and gold in the West. As settlers fanned out across the continent, they moved farther and farther from coastal population centers, meaning they had to become increasingly more self-reliant with regards to their food supplies. Initially, this consisted of anything they and their neighbors could grow themselves, as is illustrated in popular media like *Little House on the Prairie*. Soon, however, networks between communities and regions evolved as Indian footpaths became trails, became roads, and became railroads. Still, the majority of food was grown, harvested, and consumed locally, particularly in rural areas.

However, unsustainable agricultural practices, such as the decimation of the topsoil of the Great Plains through deep ploughing and uprooting of native grasses, caused as much as 75 percent of the topsoil in some regions to be eroded.[5] In fact, on April 14, 1935, roughly three hundred million tons of topsoil was dislodged and became a colossal dust storm. This day became known as "Black Sunday" and saw more dirt blown across the country by the wind than was moved to build the Panama Canal.[6] The Dust Bowl became emblematic of Americans' failed relationship to their land. The dust clouds that billowed Eastward as far as New York City made it clear the promise of modern agriculture meant nothing if it was not coupled with environmental stewardship.

All this began to change with the massive economic disruptions of two World Wars and the Great Depression. The Great War catapulted the US to global agricultural stardom, having to supply the Allied nations and their armies with much of their food supply. This supported the economic ex-

---

5   Richard Hornbeck, "The Enduring Impact of the American Dust Bowl: Short- and Long-Run Adjustments to Environmental Catastrophe," *American Economic Review* 102 no. 4 (June 2012): 1477-1507.

6   Oklahoma State Legislature, Senate, "A Concurrent Resolution Designating April 14, 2015 as 'Dust Bowl Remembrance Day,'" S 16, 55th Oklahoma Senate, 1st sess., adopted in Senate April 13, 2015.

pansion of the Roaring Twenties, but left farms vulnerable for the dearth of demand during the Great Depression. In the depths of the economic tumult that defined the 1930s, the small farms which formed the backbone of the American food system began to crumble in earnest. Prices bottomed out, causing many farmers to declare bankruptcy. Corn prices fell to just eight to ten cents per bushel, causing farmers to burn their corn in their furnaces instead of more expensive coal, perfuming the countryside with the smell of popcorn.[7]

A dust storm blows through a small Texas community in 1935.

President Theodore Roosevelt believed the key to America's economy was supporting the rural economy of America's heartland. To this end, his New Deal included two provi-

---

7   "The Great Depression Hits Farms and Cities in the 1930s," Iowa
    Public Broadcast Corporation, accessed August 22, 2020

sions to bolster American food production. In 1933, the Agricultural Adjustment Administration (AAA) sought to raise prices through shrinking surpluses by paying farmers not to plant on some of their lands. In 1937, the Farm Security Administration (FSA) was created to combat rural poverty with intensive land conservation programs and even a resettlement scheme for farmers whose farms were compromised by the Dust Bowl.

But as World War II began, FDR realized the war effort had to be won also in the urban core. In the first exercise of treating locally sourced food as a policy priority in US history, Victory Gardens were established across the country in front yards, vacant lots, public parks, and anywhere in between. In addition to the "moral win" of letting those on the home front feel they were contributing to the war effort, the program resulted in over five million new gardens and netted over $1.2 billion in farm goods.[8]

Unfortunately, these accomplishments were discarded in favor of the mass industrialization in the post-war era.

Mass Industrialization

After the war was won, a major chunk of the war economy was redirected to the agricultural industry. Ammonia plants made for warfare were repurposed to create nitrogen fertilizer for farms. Factories that once created engines for

---

8   Charles Lathrop Pack, *War Gardens Victorious* (Philadelphia: J. B. Lippincott, 1919), 15.

tanks began to manufacture them for tractors and new cotton-picking machines.

Suddenly, the need for manpower on the farm was depleted and a mass urbanization led rural men and women to the cities and suburbs. As suburbs expanded, they pushed farms farther and farther from their urban consumers. New government policies such as farm loans, encouragement of monoculture and industrialization in the 1970s, and our modern system of farm subsidies in the 1990s drove more and more farmland into corporate hands. There was no doubt the government had prioritized farm profits and efficiency over all else.

The federal government's message to farmers was simple and is best expressed in the words of Richard Nixon's Secretary of Agriculture Earl Butz: "Get big or get out!"

These massive farms used intensive irrigation, pesticides and fertilizers, and giant machines to produce crop yields the likes of which the world had never seen. Of course, what comes up must come down. This fantastical harvest was not coming without a price to pay.

Agents for Change

As the topsoil depletes, groundwater becomes toxic, and the aquifers run dry, it's no wonder why the modern environmental movement is wary of factory farms shipping produce around the country and the world.

Looking ahead, this system isn't working for our children, many of whom have become very disconnected from where their food comes from. It's no secret by and large, American kids have a poor impression of produce. They'd much rather munch on a Big Mac than a brussels sprout. All in all, most ordinary people are alienated from the production of their own food and are not economically invested in it.

Plus, our economy is structured to support big agribusiness and other food giants over small, locally owned businesses.

With the state of our food system this way, it felt disheartening to wish for a time before corporate control of food production. It was only a matter of time before people began to fight back. In 1986, in protest of the construction of a McDonalds near Rome's iconic Spanish Steps, political activist Carlo Petrini founded Arcigola, an organization dedicated to reforming organic relationships between people, food, and culture. Now Slow Food International, the group promotes local cooking and agriculture using native seeds.

Revolutionary at the time, SFI is now in over 150 countries worldwide. In 2005, San Francisco chef Jessica Prentice coined the term "locavore" to describe the growing movement of folks tying their diets to their local "foodsheds" of one hundred miles or less. Now the word is common parlance for anyone partaking in their community's food bounty. Politically, this has evolved into the broader conception of "food sovereignty," or the idea communities should have control over their own food supply. The state of the food sovereignty movement is still small, but already sprouts of hope are beginning to shoot up between the cracks.

Of course, these small changes cannot solve our manifold problems: environmental degradation, widening economic inequality, worsening health outcomes, and more—we need a bold, new approach to tackle the impending trials of the twenty-first century. A farmers market in each neighborhood would not be enough, but rather, we must rebuild regional food networks and create local food sovereignty in communities across the country.

Such an approach would go a long way for communities and investors to reclaim the economic value in their diets. It would support local producers who would strengthen local economies and put money directly back into their own regions. It would build resilient, local food systems which can take on the challenges of climate change and national security threats like pandemics more nimbly than a global supply chain. It would help reintroduce heritage seeds and promote regenerative agricultural practices. It would promote nutritious, fresher diets to build the next generation of healthy Americans. This notion is perhaps best expressed by Karen Washington, a New York-based food activist and founder of Rise and Root Farms:

"We're all suffering. But at the end of the day, folks, what makes us strong is our belief in one another, that we will come together to help one another get back on our feet... This is our time, this is our moment to not go back to politics and Wall Street, but to move forward. It's more about people than profits. This is our time to move forward and change the system."

All in all, it would help the United States to build a more robust, healthy, environmentally friendly, profitable, equitable, just, resilient, and tastier food system for this century and beyond.

CHAPTER TWO

LITTLE FARM, BIG IMPACT: NATURAL-SCALE LOCAL AGRICULTURE

Marbury Jacobs and Taylor Yowell stood on their back porch and looked out over their land. The sweltering humidity and searing sun of the Mississippi summer morning weighed heavy on the back of their necks as they set out to begin their weekly ritual of picking, washing, and packaging their crops. No machinery, no pesticides, no herbicides, no plastics, and no farmhands—just a happy couple harvesting everything from staple vegetables, like okra and spinach, to cut flowers, and to their own signature tea blend. When the hard field work of the day was done and the midday sun began its rise high above trees that ring their homestead, they loaded the bags bursting with fresh produce into the back of Taylor's truck. Later that day, he'd drive into Jackson and drop off these "veggie shares" to their eager customers.

Little Farms Mean Big Business

While this little vignette may evoke sentimental feelings of simpler times in the agricultural South, The Garden Farmacy, LLC, Taylor and Marbury's plot, is very much a part of the modern food system in the Jackson metropolitan area. On just six acres, they feed over one hundred local families every week during the growing season. In a state dominated by massive corporate farms growing commodities like cotton for export, to textile mills in East Asia, and soybeans for use in animal feed, The Garden Farmacy, as a "natural-scale" farm specializing in fruits, herbs, and vegetables, is an anomaly. Mississippi, although blessed with some of the world's most fertile farmland (one need only to crack open any American high school history textbook and read about the massive plantations of the Mississippi Delta) imports about 90 percent of its food every year. Marbury and Taylor are making a dent in that massive number on their own small plot of land and they're doing it the natural way with a modern, innovative approach.

Instead of selling their harvest to a big supermarket or a distributor, where it may travel hundreds of miles and sit around in trucks and on shelves for weeks before being consumed, they utilize a Community-Supported Agriculture, or CSA, system. Increasingly popular amongst small farms across the United States, this model allows local consumers to purchase "shares" in a certain farm's harvest for a season. In return, they receive regular deliveries of fresh, local produce directly from the farmers. This model allows for the community to use their purchasing power to support their agricultural neighbors while also consuming local produce that is fresher, tastier, more healthful, and environmentally friendly. Over 12,500 American farms participate in CSA programs and the number continues to grow as more folks begin to care about where their food comes from.[9] Americans everywhere are returning to local methods of food distribution, whether through CSAs, farmers markets, green grocers, bulk sales to restaurants, direct-to-consumer online orders, and even conventional supermarkets sourcing locally. This helps to weave the web of an alternative, decentralized food system that does not rely on long supply chains and the yields of corporate farms thousands of miles away.

Going "Beyond Organic"

Many farms in the United States, especially massive farms that care little about their environmental impact, are caught in a deadly cycle of corporate dependence. They are sold genetically modified seeds from large seed companies, such as the infamous Monsanto agrochemical company, with the promise their yields will increase dramatically. These seeds will produce hardier crops that grow faster, need less water, and bear larger fruits. However, there's a catch: For these seeds to produce the intended results, farmers need to buy expensive fertilizers and harsh pesticides, also from big agribusinesses. The pesticides definitely kill off pests, but they'll also kill off the beneficial insects and bacteria plants and the earth need to keep soil healthy. This is where the intense fertilizers come in

---

9   U.S. Department of Agriculture, National Agriculture Statistics Service: *2007 Census of Agriculture* (Washington, D.C., 2009), 606.

to supplement the lost natural productivity of the earth. Those same fertilizers, along with the pesticides, mostly run off with the topsoil into waterways, contaminating local ecosystems and eventually make their way into the human diet.

As the farmland loses its vitality, it will require more trade-marked seeds and even more chemicals to keep the farm productive. In the end, big corporations make a killing and the farmers who were sold a promise of a bountiful harvest end up doing the killing. Even so-called "organic pesticides," such as those which employ copper compounds, are still synthetic and often toxic. This, along with the prohibitively expensive and time-consuming process which the Department of Agriculture mandates to achieve an organic certification, has caused many small farms to forgo the organic label entirely. Instead, many farmers have disavowed the official USDA "Certified Organic" label and instead have opted for what they refer to as "beyond organic." Marbury Jacobs of The Garden Farmacy says this of her farm's embrace of the concept:

"Ultimately, it comes down to the trust of our customers. They know us. There's the saying: 'Know your farmer, know your food.' So, our customers trust that we're not spraying anything [and the reason] we don't need to spray kind of comes down to this fundamental understanding of the ecology of the farm. We're not in pursuit of a label or a marketing tactic. It's more of a way of life for us and our farm. We strive for nature to play as much of a role as possible rather than trying to create a sterile or artificial growing environment."

Taylor Yowell and Marbury Jacobs on the fields of The Garden Farmacy in Bolton, Mississippi.

Small farmers who wish to farm "beyond organic" will concentrate on improving soil quality, thereby developing a microbiome (that is, a community of microorganisms in a particular area) on their farm that will naturally support a healthy ecosystem and therefore healthier crops. This can be done by practicing polyculture, that is, by strategically planting a variety of plants close together that complement each other within their ecosystem. These plants may have different functions in the ecosystem and will interact with each other and the bacteria in the soil in different, positive ways as well as the other organisms who call the farm home. Some may fix nitrogen levels while others may provide habitats for beneficial pest-eating insects like ladybugs. Some may act as windbreaks or canopies to protect smaller plants, and some provide ground cover to protect soil not being actively cultivated. Plants serve many more functions, and when thoughtfully grown together, they collaborate to produce a strong, healthy microbiome that is in harmony with the ecosystem in which it grows.

When polyculture is coupled with diligent mulching to block out weeds, compost to enrich the soil with a panoply of nutrients, and years of proper, conscientious management will yield crops that are not just more resilient to pests and climactic conditions, but also more nutrient-dense and delicious. Thus, small farms can ignore the time-intensive and costly organic labelling process and spurn the corporate racket of designer seeds, toxic pesticides, and chemical fertilizers in favor of time-honored methods of agriculture that work alongside the natural ecosystem. The result is farmers save valuable money and natural resources which they can use to put their business on more stable financial footing, preserve the ecology of their farm, and to serve even more members of their communities.

The Deindustrialization of Agriculture

In addition to exploring "beyond organic" agricultural methods, a feature that only small-to-medium-scale farms can employ is deindustrialization. Businesses like The Garden Farmacy don't own a tractor, combine harvester, or any other heavy machinery to work the land. Instead, they work at the human scale—tilling their soil, planting their seeds, watering their crops, and harvesting their produce all by hand. In our modern world, this may seem not just counterintuitive but anachronistic. We have been so thoroughly indoctrinated into the cult of progress the mechanization and automation of every task possible seems to be the only desirable result, and why wouldn't it be? Don't machines take the work out of farming? The truth is, while corporate farms can only be run with the help of technology, on a farm that may be just a few acres (or not even one acre), machines are not a necessity. Especially with plots on the smaller end of that scale, it's important to use as much of the land for cultivation as possible. Ergo, the amount of physical and financial space a tractor would take up, with rows of crops being replaced with paths for the tire treads, an on-site mechanic on standby for the perennial mechanical issues, and the cost of maintenance and gasoline just does not make sense.

Small farms can save money while saving their local air from combustion engine pollutants. Not only that, but the hand tools often do a more thorough job than the machines. For example, tilling the soil with hand implements allows the farmer to achieve a deeper depth of tillage than would typically be possible with motorized cultivators, permitting greater aeration of the soil and helping perennial crop roots access deeper down closer to the subsoil. The benefits of this method are two-fold: plants become more firmly rooted in the ground and less susceptible to weather events that might uproot shallower plants. Also, this lets plants access water stored deeper underground, a bonus that is especially beneficial during the summer months. The Garden Farmacy, for example, makes use of hand tillage for their crops so they don't have to irrigate their fields at all, even during the long, dry summers of the Deep South, saving time, money, and the local environment.

Farming Small on the Hawaiian Islands

Of course, the cynics might say while the thought of buying most of our produce locally is nice, it's just unrealistic. It's simply impossible for local small farmers to compete with the prices and scale of huge plantations hundreds or even thousands of miles away. That's where food hubs come into play. A food hub is an organization that helps to aggregate a region's local food products to market and sell them to grocery stores and restaurants. With the mission of supporting a region's small, independent producers and helping them to compete with large, often out-of-state commercial farms, food hubs are vital in bolstering a region's ability to feed itself in an economically viable way. While food hubs have sprung up around the country, one of the most poignant examples of their positive affects is Maui Food Hubs on the Hawaiian island of Maui, a food hub dedicated to bolstering the autochthonous food systems that have nourished Hawaiians for generations.

Native Hawaiians were able to produce an abundance of food to feed themselves before colonization due to the rich, volcanic soil for which the islands are known. Growing native Southeast Asian and Polynesian crops like taro, breadfruit, and kukui nuts, Hawaiians practice the concept of "Aloha 'aina," which translates to "love of the land". This saying speaks to the love and respect they have for their land as a sustainer of life. The Hawaiian word "kuleana" also speaks to a shared responsibility to their environment and their communities. The islands were divided into a traditional subdivision called "ahupua'a," which usually extended between ridges from the top of mountains all the way to the ocean and further divided into extended family plots called "'ili". This organization encouraged families to farm along the waterways which flowed down the valleys and through their fields, carrying nutrients with it from 'ili to 'ili. This natural irrigation nourished crops by harnessing the power of run-off and also helped bring nutrients into near-shore marine ecosystems. However, the American conquest of Hawaii, which was a coup supported by the white sugarcane and pineapple plantation owners, paved the way for corporate monopolies to commandeer large swaths of the most fertile land across the archipelago. These colonizers sapped the land, rerouted heritage waterways, and destroyed native foodways with monoculture fields of foreign pineapple, macadamia, corn (now the most common crop on the islands), and more. Corporations like Monsanto bought up large parts of the flatlands to harness Hawaii's three growing seasons for the testing of new genetically modified crops, turning the islands into a laboratory. Now with over 1.4 million inhabitants, Hawaii is the only major population center in the Pacific Ocean and they must import the vast majority of their food at steep prices to both their wallets and the environment. Interestingly enough, when Hawaii was first visited by Captain James Cook in 1778, there were already roughly one million people living on the Hawaiian Islands.[10] If they were able to achieve self-sufficiency in food production before the American annexation, could they not reclaim it again?

This is where native Hawaiian farmers and other stewards of Hawaii's farmland, dedicated to reviving local agriculture and the organizations that support them, come into play: Launched in response to the COVID-19 crisis, Maui Food Hubs is trying to change the way Hawaii eats by connecting producers and consumers on the island of Maui. On their website, one can browse local farmers, fill up a cart with their products, and pick it up every week at a drop-off location nearby. Maui Food Hubs has come at a critical time as COVID-19 has interrupted the long-standing food supply lines that have kept Hawaii fed since annexation. Plus, a steep decline in tourism, the state's largest industry, has left locals who rely on visitors for their paychecks high and dry, including the farmers who provide the food for hotels and resorts. Meanwhile, the profits from the industry largely go into the hands of mainland corporations and a small cadre of extremely wealthy executives, neither of whom are feeling particularly generous with regards to ameliorating the food crisis. This predicament has forced the islands to reckon with their brittle food system, a discussion in which

---

10 David E. Stannard, *Before the Horror: The Population of Hawai'i on the Eve of Western Contact* (Honolulu: Social Science Research Institute, University of Hawaii, 1989), 66-67.

Maui Food Hubs is leading the way. They have successfully lobbied the government of Maui to convert the 2020 tourism marketing budget into a micro-grant program for farmers to ramp up production through the purchase of irrigation systems, machinery, and more. They've also found if Maui Food Hubs handles the work of weighing, cataloguing, packaging, and distributing the produce, small farmers have more time to care about their crops and improve their yields. Some of their clients have estimated they will grow anywhere from 30 percent to 50 percent more food with this arrangement. This model has incredible potential to buoy Hawaii's food output and feed more and more people without resorting to imports.

The local community has also embraced the program, with lines around the block every weekend on pickup day. There is also trepidation about what happens post-COVID when the tourists return to the islands. Autumn Ness, one of Hawaii's most fierce advocates for small farmers who helped build Maui Food Hubs, expressed the joy and frustration of the local population:

"The community is asking us to please not go anywhere. They didn't know what kind of beautiful produce we had on Maui because they were all going to the tourists. They say wow, I've never seen lettuce and tomatoes like that! It's actually infuriating that locals are going to the grocery store and paying ridiculous prices for head of lettuce that's been on a boat and through the whole North American food system before it gets here... So, what we're seeing is that the tourists have been enjoying all this beautiful organic local produce and only now is the community seeing it."

The crisis has also laid bare the stark racial disparities of the Aloha State. The lines for the fresh produce boxes were almost entirely white, while just two blocks away, the line at the food box giveaway at a local elementary school, which was giving out boxes filled with Spam, macaroni and cheese, and assorted canned goods, was nearly all Native Hawaiian. Maui Food Hubs hopes to begin to address this inequality by accepting SNAP, WIC, and other supplemental nutrition programs and by implementing a "pay it forward" option for customers to buy boxes for families in need. However, they understand broad systemic change is the only way to rectify the situation. Autumn says it best:

"The hotels are owned by off-island interests. They employ locals at $10 an hour, all the profits go somewhere else, and all those people are the ones that got immediately laid off. So, we're at 40 percent unemployment right now because of COVID and almost all of those people are locally-born, whether they're descendants of plantation laborers or they're native Hawaiians... Hawaii is like a magnified version of the failure of capitalism and colonialism in our food system. So many people just don't get that in the US."

Where our food comes from is just a part of the equation. Just like the Hawaiians, residents of all fifty states and territories must also consider who grows their food, how it is grown, where it ends up, and who profits from it. If we can put our food systems more squarely into our neighbors' trusted hands, we can start to reimagine the food landscape of our entire community. By empowering regional producers and distributors, we will improve our environment, stabilize local economies, and make local residents happier. After all, why rely on others for what you can do for yourself?

CHAPTER THREE

A FARM ON EVERY CORNER: AGRICULTURE IN THE BIG CITY

According to the Food and Agriculture Organization of the United Nations, roughly eight hundred million people around the world are growing produce or raising livestock in the urban environment as of 2015. This accounts for nearly 20 percent of the world's food![11] In the US, the last iteration of the Farm Bill, a massive omnibus bill that directs the federal government's food and agricultural policy, featured the very first supportive legislation directed specifically at urban agriculture in 2018.

---

11 *The Place of Urban and Peri-Urban Agriculture (UPA) in National Food Security Programs* (Rome, Food and Agriculture Organization of the United Nations, 2011).

The State of Urban Farming Today

Nowadays, people can find food growing on parts of the urban environment that have traditionally been underutilized—places like rooftops, vacant lots, the walls of apartments and office buildings, and even in underground tunnels! They can be high-tech operations with robotics and hundreds of data points or just a group of people seeing what'll grow on land that used to be a parking lot. Since many urban farms are growing on a very small scale (think less than one acre) they tend to focus on a variety of crops that are high in value, compact in size, and can be purchased wholesale and used at the hyperlocal level. Oftentimes, this manifests itself in the form of the perennially trendy microgreens. These tasty infant versions are picked at the sweet spot between two to three weeks of growth and may be yielded from common edible plants like amaranth, sunflower, and arugula. They are prized by chefs and foodies alike for their bright color, crisp taste, and claimed nutritional value. Small and delicate, they are perfectly suited for the hyperlocal market which ensure chefs and home cooks can get the most out of their short shelf life.

Since its inception, urban agriculture, especially indoor operations, has had an upper bound on its profitability due to its high overhead costs. Smaller harvests, high land value and property taxes, pricey high-end farming equipment, expensive utility costs, and more have long confined urban farms to the realm of specialty produce. However, larger urban farming operations have expanded out of the specialty produce space and into more traditional crops as they've scaled their production up and kept their overhead costs down. For example, AeroFarms, the massive, tech-savvy urban farm explored in the introduction, can sell produce like lettuce at economical prices to local supermarkets, competing directly with typical farms. Their innovative vertical farming technology uses a system called aeroponics, where they replace soil with a reusable cloth growing medium and spray the plants' roots directly with a mist of water and all necessary nutrients. Along with tailored LED lighting (the cost of which has dramatically fallen in the past decade), a high-tech circulation system that can recycle the nutrient-rich vapor the plants do not absorb, and data analytics for every harvest, their system has resulted in yields that are nearly four hundred times more productive per square foot than field agriculture while using 90 percent less water. Groundbreaking technology like this is what will propel urban agriculture into its next great evolution.

Another fabulous benefit of indoor agriculture is its capacity for tailoring climate conditions perfectly for each plant. For example, 80 Acres Farms, an indoor vertical agriculture operation in Southwestern Ohio, is able to toggle temperature, humidity, light, airflow, carbon dioxide, and nutrients to create the exact optimal growing conditions for each varietal. The result are plants that grow on average three times faster than their outdoor counterparts, yielding harvests one hundred times larger. In fact, their name comes from the fact that in their twelve thousand square-foot combined growing spaces, a little over a quarter of an acre, they are able to grow eighty acres worth of produce! This sort of efficiency could only be possible through the technological prowess of an advanced indoor growing operation and cannot be replicated in a traditional field setting.

Moreover, because of its sterile indoor environment, no pesticides ever have to be employed, keeping the produce as fresh and healthy as possible. However, the rare times some pests do break into the facility, 80 Acres employs an ingenious strategy taken straight from nature's playbook. Liz Warren-Novick of 80 Acres Farms explains:

"In order to counteract the bad bugs, we actually have a biologist named Alex who knows everything there is to know about different insects. If a 'bad bug' were to somehow get into one of our growth zones, he would know which 'good bug' to counteract it with. So, he gets to kind of create his own little ecosystem and study how everything plays out. So, since we don't use pesticides, we utilize ladybugs, nematodes, rove beetles, and spiders."

Inside the basil "Grow Zone" at 80 Acres Farms in Hamilton, Ohio

In this way, 80 Acres uses science to improve upon nature's work and unleash the potential in every seed. This combination of cutting-edge technology and a deep respect for how natural systems play out makes indoor agriculture powerful and productive.

Why Would Anyone Want a Farm on Their Corner?

Along with the tremendous value in farming urban and nontraditional environments, the positive externalities are profound. Environmentally, most indoor agricultural systems use dramatically less water than their outdoor competitors. By targeting their plants more precisely with innovative irrigation systems, indoor farms are eliminating runoff and lowering their water bill. Soilless farms making use of systems like hydroponics and aeroponics are choosing not to make use of our diminishing supply of topsoil. Urban farms that do use soil often provide and make use of compost generated by the city, helping to curb food waste. Also, seeing as urban farms are significantly closer to their consumers than their rural counterparts, environmental costs associated with transportation are considerably diminished.

On the health front, the benefits of having fresher produce is obvious: produce which has spent less time between field and plate has spent less time rotting. It will not just taste better, but also retain more of its fiber, vitamins, minerals, antioxidants, and other phytochemicals (chemicals that naturally occur in plants). Diets replete with fresh produce save humans from a variety of maladies associated with high consumption of refined sugar and processed foods, such as obesity, diabetes, heart disease, and tooth decay. What's more, eating plenty of fresh produce results in better mental outcomes. Also, 95 percent of the serotonin (a neurotransmitter that promotes feelings of happiness) your body produces originates in your gastrointestinal tract.[12] A wealth of fresh fruits and vegetables in your diet promotes a gut microbiome which is conducive to the production of serotonin and limits inflammation, promoting mental well-being. As an added bonus, bringing greenery into the urban landscape is beneficial both for mental health and for productivity.[13] The farm you pass on your walk to work just might make you a happier person and a better employee!

Economically, urban agriculture is a windfall in terms of local employment and retaining dollars in the community. Many traditional agricultural jobs are seasonal—think of the famous grape farmers who César Chávez and Dolores Huerta organized in the 1960s. Workers came and went with the harvest season, forcing them into unstable, low-paying jobs and an itinerant lifestyle. However, indoor agriculture, with its year-round growing season due to climate control and its incorporation of technology and data, creates permanent employment opportunities for individuals from a variety of

---

12 Natalie Terry and Kara Gross Margolis, "Serotonergic Mechanisms Regulating the GI Tract: Experimental Evidence and Therapeutic Relevance," *Gastroinestinal Pharmacology Handbook of Experimental Pharmacology*, 239 (2016): 319-342.

13 Anne C. Bellows, "Heath Benefits of Urban Agriculture Public Health and Food Security," Food Security Dot Org, accessed June 20, 2020.

educational backgrounds. AeroFarms has 85 percent of their workforce living within fifteen miles of their Newark headquarters. CMO Marc Oshima can see first-hand the effects they've had on their employees:

"We decided to interview a couple of our team members to give them a chance to tell their stories...in one particular case, a team member was able to buy a home by having the financial security of a job. He never imagined working in agriculture. He grew up in Newark... We had another team member who was in the financial position where she could bring her mom over from another country to bring the family together for the first time in thirty years..."

It's clear meaningful, stable employment in the urban agriculture sector is capable of changing lives for the better. On the macro level, having farms close by allows for the money from household produce budgets to stay in the area. Buying from your neighbors instead of corporations increases the velocity of the dollar in your area, meaning dollars are spent more often and move around an economy faster rather than collect dust in a corporate bank account. This means each dollar works harder to create economic output in your region, the effects of which multiply throughout the local community, catalyzing economic development.

In the realm of urban development, urban farms are notorious sparkplugs for neighborhood renewal. They take over empty lots and disused property and improve the value of the land on which they grow and the lots nearby. They often make neighborhoods more appealing places to live (as long as you don't mind the smell of manure!) and even have a positive effect on crime.[14] This is perhaps most poignant in Detroit, the American posterchild of urban decay.

Turning Motors into Tillers in Detroit

In a city marred by capital flight, the legacy of institutionalized racism, and the loss of much of their foundational automotive industry, Detroiters are beginning to write a new chapter for their hometown through urban agriculture. Blocks of urban decay have given way to verdant fields as over 1,500 farms and gardens have sprung up across Motor City. One of the smaller ones is Fisheye Farms, started in the parking lot next to a dry cleaner in the Indian Village neighborhood of the East side of Detroit. The founders of Fisheye Farms, Amy Eckert and Detroit native Andy Chae, realized ten thousand dollars of revenue in their first year selling to folks within a few blocks of their farm. Seeing an opportunity, they quickly began to expand, buying land in the nearby city of Pontiac and 0.89 acres worth of vacant plots in the urban core of Detroit. Their small-scale commercial operation now offers transplant sales to other local farmers, wholesale produce to neighborhood restaurants, a direct-to-consumer farm stand, and a seasonal produce box subscription.

Amy and Andy's success on their bottom line isn't the only reason Fisheye Farms is making a positive difference in Detroit. Their neighbors in Indian Village have welcomed

---

14 "Breaking Barriers: How Urban Gardens Impact Crime," Abundance North Carolina, accessed June 20, 2020.

them with open arms. Amy cannot overstate how their business has created space for their local community to bond, build trust, and share resources:

"When we initially were interested in buying the land, we went to the houses on the block and told them a little bit about what we were planning on doing and just tried to gauge people's reactions. Everyone was a little confused and probably pretty skeptical but still liked the idea. As we started putting in time at the space, we slowly met everyone and found out what their interests in the farm were. There are a lot of kids on our block and they are naturally curious about everything. As we met the kids we would get introduced to their parents and get to know them more. One thing that really brought us close to a few families was having our tool storage shipping containers broken into and equipment stolen. They all felt really bad for us and now do everything they can to watch out for us, even calling us if they see someone snooping around at night. We let them borrow tools, let some folks who don't have electricity charge their phones, give them fresh produce, and give their kids a safe place to hang out and explore. Our neighbors all came to our wedding at the farm almost two years ago. We're all just friends and a great small community!"

Stories like this are a dime a dozen when it comes to urban growing. It turns out farms make great neighbors, and when you create productive, welcoming spaces that benefit your local community, the community in turn will benefit you. As our cities continue to burgeon with opportunity and neighborhoods go through demographic and socioeconomic changes, urban farms can provide a stable, safe space where community members can still come together around fresh food and fresh air. Beyond the tangible positives discussed earlier, studies have found the introduction of urban farms into communities "offered multiple perceived benefits including increased social connectedness, a transformed physical landscape, improved neighborhood reputation, increased access to fresh produce, and educational, youth development, and employment opportunities."[15] Clearly, urban farms offer more than a tomato grown down the street—they become economic, cultural, and social hearths of their entire area. To this end, Fisheye Farms is one line in the story of Detroit's agricultural revolution—providing healthful, local fruits and vegetables to the people of Indian Village while uplifting their local economy and sense of community.

Urban agriculture brings the farm to our doorstep and in the process, the bounty of positive externalities they offer to your street, neighborhood, and city. While one urban farm by itself may not be revolutionary, a massive citywide network turns the region into a food-producing behemoth and integrates farming into the culture and urban fabric of the community. This is the real work of the urban farming movement—bringing an industry that was once seen as the sole property of rural folks and making it relevant to the lives of the urban denizens who crave a closer connection to their food.

---

15 Melissa Poulsen, Roni Neff, and Peter Winch, "The multifunctionality of urban farming: perceived benefits for neighbourhood improvement," *Local Environment* 22 no. 11 (2017): 1411-1427.

CHAPTER FOUR

THE ART OF THE VEAL: BRINGING LIVESTOCK HOME

Balancing land management with agriculture has always been a challenge and governments around the world have struggled to address the issue. In the late 1960s, the colonial government of Rhodesia was faced with an encroaching desert and a retreating grassland as hundreds upon hundreds of acres of once-productive land became worthless every year. Uncertain as to why this was happening, the authorities were ready to do anything to stop the advance of the wastelands. They decided to take the advice of Allan Savory, a Game Officer in the Northern provinces, and do something we would now consider heinous: the wholesale slaughter of elephants. Convinced the herds of elephants were trampling and destroying their own habitat, the Rhodesians killed roughly forty thousand elephants throughout the early 1970s. Predictably, this did not work. Savory was relieved of his job and left the country. But his fascination with how grazing animals and their habitats interacted only grew.

The Big, Meaty Picture

As heinous and nonsensical as the largescale slaughtering of elephants to fight desertification may seem, this thought process is actually characteristic of a prevailing mindset in modern agriculture: having too many animals in one area will necessarily lead to overgrazing and the degradation of the land. It's something meat-eaters across the country have heard time and time again—eating meat and dairy is the single worst thing one can do to contribute to climate change. It seems that every day, more and more Americans are going vegan and have deemed meat consumption to be incompatible with their modern environmental sensibilities. Surveys show about 6.5 million Americans identify as vegans and one in twenty Americans are vegetarians.[16] But the fact is raising livestock and eating meat itself is not the problem, but rather the issues stem from how our modern economy and industrial agriculture has warped the sector into an ecological disaster. Fortunately, we have ways to not only lessen the environmental impact of meat production but to make meat production a net positive for our communities and our planet.

In the United States today, meat production is the single largest sector of our agricultural economy. This huge industry receives thirty-eight billion dollars in annual subsidies from the federal government (compare that to just seventeen million dollars for fruits and vegetables).[17]

---

16   Charles Stahler, "How Many People Are Vegan? How Many People Eat Vegan When Eating Out? Asks the Vegetarian Resource Center," date accessed April 18, 2020.

17   David Robinson Simon, *Meatonomics: How the Rigged Economics of Meat and Dairy Make You Consume Too Much* (Newburyport: Conari Press, 2013), xxi.

The meat industry is also notoriously toxic to our environment. Many industrial livestock farms in the United States employ a model known as CAFOs, or concentrated animal feeding operations. In a CAFO, thousands of animals are crammed into massive sheds where they are kept in extremely close quarters, sometimes not even able to turn around. They are often bred to grow fast, many of them buckling under their own weight and developing chronic pain, sores, and infections. Many never get to leave the facility except for slaughter. The goal here is not to produce healthy, happy animals, but instead ruthless efficiency.

While CAFOs and similar industrial practices have kept the price of meat at record lows, we all pay for it at the cost of our health and our environment. CAFOs produce millions of tons of manure and other animal wastes which can poison local water sources, pollute the air, and spread zoonotic diseases such as e. coli, MRSA, and mad cow. While most animals in CAFO's are chickens, ruminants—animals that regurgitate their food to chew and digest it multiple times, such as cows, goats, and sheep—are the poster children for livestock's effects on climate change. Global ruminant animal farming produces incredible amounts of methane, a greenhouse gas that is twenty-three times more effective at trapping the sun's heat than carbon dioxide. All in all, it's no surprise why people are critical of meat and dairy consumption. With global demand for meat rising every year, we need to make some hard choices about the direction of this industry.

Raising Cattle and Healing the Earth

After the disaster with the elephant massacres, Allan Savory became obsessed with how grazing animals can benefit degraded land. In the 1970s, after studying the ecosystems of grasslands around the world, Savory posited that ranchers could use their livestock to mimic the natural herds who once roamed the prairies, steppes, and savannas of the world. Through the careful implementation of proactive agricultural practices with an eye toward the water, carbon, and mineral cycles, Savory found herds of cows, sheep, and goats could replace lost keystone species like bison and elephants, bolster the health of native grasslands, and revitalize entire ecosystems. He later moved to the United States where he founded Holistic Management International, an organization dedicated to promoting his new theory of land management. Today, more than thirty million acres of land around the world employ Savory's methods to make use of their livestock to have a positive impact on their local environments and communities.

Eight hundred of those acres are being managed by the folks over at Ayrshire Farm in Upperville, Virginia. Situated on the grounds of a historic working plantation, their vertically integrated business includes a farm-to-table restaurant, a farm store, a pet food operation, a boutique slaughterhouse, and their own processing facility. They have revitalized the property and have committed to responsibly and organically managing the land in conjunction with endangered heritage-breed livestock and heirloom fruits and vegetables. In nature, animals and plants comingle all the time, so it should be no surprise our farms can benefit from the same interactions. Rather than keeping nature, crops, and livestock rigidly separated, allowing them all to live together promotes a much healthier, more robust ecosystem. The plants benefit from the nutrient-rich animal waste, as well as their natural grazing that promotes deeper root systems. This also sequesters a tremendous amount of carbon as root systems burrow deep into the soil. As a certified Predator Friendly farm, livestock are free to roam away from their paddocks and into the wooded areas and natural fields around the farm where predatory species also live. While it may seem counterintuitive to potentially put your livestock in harm's way, Alexis Russell, a Compliance Officer with Ayrshire Farm, explains how they purposely maintain natural spaces to mitigate the issue:

"So we recognize that we want to be profitable, but we also understand that we're farming with the environment. We have a ton of wooded areas on our farms where our pigs like to hang out. It's also where our local bears like to hang out. But because we have a great biodiversity load on our farm, we don't have a lot of predator loss. They don't need to try to break through the electric netting to get to the chickens because they have all the rabbits in the woods that they could possibly want to eat. So, we just don't have to deal with it that much."

In return, a lot of wildlife benefits from proximity to the farm, such as the growth of native grasses in pastures and native endangered bird species utilizing those pastures for nesting. Ayrshire, for example, supports habitats for four out of the five of Virginia's critically endangered bird species in its pastures and works with nonprofit and government groups to count bird populations and help conserve the habitat. In this way, the farm is not apart from the landscape, but is in fact a "working landscape" in its own right. The farm is not removed from the ecosystem in which it finds itself but is rather a productive and beneficial partner. It's no wonder Ayrshire estimates that, unlike most cattle farms, they may be carbon neutral or even *carbon negative*!

Their commitment to their local environment is bolstered by raising a variety of heritage cattle breeds. While fast food companies like McDonald's have done a great job at pushing Angus cattle as the "gold standard" of beef, the truth is their meat is no better or worse than many other breeds. While Angus cattle have been bred to produce a lot of meat per cow, the breed is prone to infections (thereby requiring a lot of antibiotics) and do not perform well on open pastures as opposed to indoor industrial facilities. The cattle breeds Ayrshire Farm raises, like Highland Cattle and Ancient White Park Cattle, are better suited to Virginia's climate and to an outdoor lifestyle. If ranches around the country would use cattle breeds who were better suited for their climate and landscape, not only could ranchers save time and money, but local environments would be less impacted and diners could better embrace the unique flavors of their region with local beef instead of the Angus monolith.

The alternative presented by Ayrshire Farm and Holistic Management International shows ranches do not have to be the environmental boogeymen they are often made out to be. With proper techniques, such as rotational grazing plans, allowing some parts of the farm to "re-wild," and raising cattle compatible with local ecosystems, farms can dramatically lessen their environmental footprint and potentially save money. Communities that have farms which raise livestock should be required or at least incentivized to retool and practice regenerative techniques so their operations do not degrade community health or local ecosystems. The USDA along with local and state governments can be leaders in providing zero-interest loans or grants for this purpose and to promote holistic land management principles.

Chickens and Rabbits and Honeybees—Oh My!

It goes without saying that CAFOs have overstayed their welcome as a mainstay of corporate agriculture and must be abolished. If the price of cheap meat is the destruction of our local watersheds, soil and air quality, and human health, there is nothing cheap about it. A saner economics would require those costs to be internalized by forcing meat producers to be responsible community partners and stewards of their land. A more rational approach would be to redirect the billions of dollars that are showered on massive corporate meat and dairy operations toward supporting those small producers who take care of their land, possibly at the cost of a reduced herd size.

Changing the infrastructure of the American meat sector will be extremely challenging. It will cost billions of dollars and require the education of hundreds of thousands of ranchers and the retraining of countless individuals whose livelihoods depend on our modern industrial model. But a plan of such a magnitude will ultimately provide future generations with benefits at an incalculable scale: improved health, cleaner environments, higher wages, and a closer relationship to their local ecosystems and communities—not to mention put a huge dent in the looming threat of climate change.

With all these systemic changes in place, it will also behoove ranchers to look toward local markets as their primary sources of revenue. This, of course, can be difficult as a lot of America's pastureland is often far-removed from the places where most of the meat and dairy is consumed. One cannot employ the same urban farming techniques with goats as one does with spinach.

However, addressing the shortfalls of ruminant livestock will not be enough to make meat and dairy part of the solution to climate change. In the United States, nine billion chickens are raised for meat and eggs every year compared to about thirty-two million cattle.[18] Chicken is the most popular meat in the country and the average American eats 293 eggs each year.[19] A key solution to making chicken farming more sustainable is to make it hyperlocal. The backyard chicken movement is gaining steam as more and more people raise their own chickens at home and harvest their own eggs so they can be a bit more autonomous from the corporate food system. This system feeds poultry diets replete with hormones and antibiotics. Chickens are relatively easy to raise, require little space, and can get by on food scraps. In addition to supplying fresh eggs, they consume

---

18 "The United States Meat Industry at a Glance," American Meat Institute, accessed April 18, 2020.

19 U.S. Department of Agriculture, National Agriculture Statistics Service, *September 2020 World Agricultural Supply and Demand Estimates* (Washington, 2020).

insect pests and their excrement can fertilize home gardens. Unfortunately, many Homeowner's Associations (HOAs), municipalities, counties, and states maintain regulations that severely limit or ban the keeping and slaughter of backyard livestock. While this may be understandable in dense cityscapes, suburban America has more than enough backyard space to house hundreds of millions of chickens. Instead of banning backyard chickens, it should be encouraged, and municipalities or community-based nonprofits should create free public programming around the practice of properly raising and slaughtering chickens.

But why stop at chickens? Other small livestock like ducks, rabbits, and even bees (although one may hesitate to call them "livestock" since bees cannot truly be domesticated) are plenty adaptable to the suburban and even low-density urban environments. With proper management and care, these animals would be no dirtier than your average dog or house cat but have the added bonus of nourishing their owners while connecting them intimately with their food. Bees are especially effective for urban environments as an apiary can be constructed on rooftops and in community gardens. Not only do they require just eight hours of work a year to yield over fifty pounds of honey, but they will also be a boon to local parks, gardens, and food forests, pollinating the flowers around the neighborhood and farther.[20] All in all, one of the biggest changes we can make to the meat and dairy industry is taking as much of it into our own hands (and indeed homes) as we can.

20 U.S. Department of Agriculture, National Agriculture Statistics Service, *Honey* (Washington, D.C., 2018).

A More Humane Slaughter

Beyond just the animals themselves, an often-vilified aspect of the industry are slaughterhouses. It's no secret that for meat to go from the pasture to your plate, the animal must be transported to a facility, stunned to prevent stress and pain, hung from a hind leg and bled out from an artery in a process known as "exsanguination," cleaned, inspected, and packaged. To the average American, the details of what happens inside slaughterhouses might seem grisly. They have been the focus of many animal rights and vegan exposé pieces, such as Paul McCartney's famous "Glass Walls" video, showcasing the horrors of modern slaughterhouse practices. It's certainly true massive industrial-scale slaughterhouses are places of intense cruelty: animals crammed into dirty facilities by the thousands and painfully bled-out and skinned while still alive and aware. The fast pace of the line can commonly result in unsanitary and dangerous conditions which promote the spread of disease and excrement while putting underpaid and overworked employees in harm's way. This is not to mention the environmental impact of slaughterhouses, which are responsible for pumping tens of millions of pounds of waste into our waterways and filing the air with pollutants. Also, most slaughterhouses are not Certified Organic or Certified Humane, so no matter how well the animals are treated on their farm, those certifications vanish the second the animals pass through the slaughterhouse doors.

Legislatures have long been aware of the issue of animal cruelty in slaughterhouses and have moved to solve it. In 1958, President Dwight Eisenhower passed the Humane Methods of Slaughter Act (HMSA), requiring facilities to take certain steps to alleviate animal pain, such as ensuring livestock are sedated or stunned prior to exsanguination. However, not only do claims of noncompliance abound every year, but the US Government Accountability Office asserts inspectors in slaughter facilities often do not enforce HMSA, refusing to stop the line or even submit a noncompliance report even if violations were recognized. This is often to protect the bottom line of the company as a stopped line can cost thousands in lost productivity. Plus, while the HMSA protects cattle, swine, sheep, horses, and mules, it does not apply to poultry, rabbits, seafood, and many other animals that constitute the majority of animals slaughtered for food in the United States. Clearly, our current system of massive industrial slaughterhouses and unenforced regulation are not helpful for either animal welfare, meat quality, or our environment. What if there was a better way to slaughter livestock that minimized suffering and maximized animal welfare?

First of all, slaughterhouses have simply become too big. It is impossible to protect animals, workers, and the planet in a facility where thousands of animals are processed every day. Instead of massive, centralized plants, the meat industry should begin to pivot toward regional networks of small-scale "boutique" slaughterhouses. Smaller facilities make better neighbors without the issues caused by the highly concentrated pollutants. They can take the time to pay attention to quality and animal welfare to ensure quick and painless slaughter. This will make organic and humane certifications easier to obtain and improve the quality of the meat. Plus, animals will no longer have to travel tens if not hundreds of miles to a slaughter facility, keeping their suffering to a minimum. This also has the bonus of lowering the carbon footprint of each animal and keeping meat as fresh as possible. Transitioning to a decentralized slaughter industry will require many more trained personnel and a new way of looking at the sector. Alexis Russel from Ayrshire Farms opines:

"If I could wave a magic wand [and solve a problem with the meat industry], it would be to sort of de-stigmatize slaughter and butchery as viable trades. Really, they shouldn't be any different than mechanics or hairdressers—it should be something that's really easy to learn, something that you could take a class for at your local community college, and it's a great career that you can potentially do anywhere and can make a lot of money with."

Having knowledgeable folks who care about their craft and are members of their local communities operating slaughterhouses can only result in more responsible practices, a smaller environmental footprint, and a higher quality product. Certainly, this will result in higher prices for meat. But if increased prices are what it takes for the industry to internalize costs and take meaningful steps toward reform, it is a sacrifice worth making.

Making Waves in the Seafood Industry

Advances in aquaculture, or the farming of aquatic life, have also come with boons to seafood productivity around the world, especially in China which produces most of the world's farmed fish. This is an urgent development as the United Nations Food and Agriculture Organization estimates 62 percent of seafood will come from farms by

2030.[21] However, traditional fish farms are extremely dirty enterprises. Crowding so many fish into small pens in rivers, lagoons, and on the open ocean creates massive amounts of excrement and other organic waste, killing marine life around the farm and exposing the fish to dangerous bacteria and diseases. Also, fish slipping out of their pens and into waterways, a phenomenon known as "escapism," is a major issue on these farms. It causes swarms of predatory fish near the farms, disrupting local ecosystems. This problem is often compounded by non-native farmed fish escaping into the waters around the farm, thereby becoming invasive and presenting a threat to native species.

Luckily, there is a better way to farm fish: new technology such as the Recirculating Aquaculture System (RAS) has allowed savvy entrepreneurs to farm fish on land in an efficient closed-loop system. These state-of-the-art systems are scalable solutions that raise healthy fish in clean environments with little need for the high doses of antibiotics which are notorious in other aquaculture operations. What's more is the waste the fish produce can be transformed into a high-quality organic fertilizer for use on local farms. Eric Pederson is the CEO of Ideal Fish, a farm which uses RAS technology to raise Mediterranean Sea Bass in Waterbury, Connecticut. At full capacity, his facility can produce about 150 tons of the coveted fish which would typically have to be raised, processed, frozen, and shipped from Southern Europe. What used to take over a week now takes less than

---

21 *Fish to 2030: Prospects for Fisheries and Aquaculture* (Washington, The World Bank, 2013).

twenty-four hours from farm to plate. Eric has seen just how pleasantly surprised people have been with the superior quality of his product:

"When we first started out, we started selling in supermarkets, and the first time the fishmonger saw my fish, he called his distributor and complained about the fish arriving frozen... Turns out, the fish wasn't frozen, but was so fresh, that it still had rigor mortis! He had never seen fish so fresh before that."

As appetites for seafood are only growing nationwide as oceanic and riverine ecosystems decline around the globe, RAS and Ideal Fish illustrate a sustainable, high-tech solution to a pescatarian's dilemma.

Facing the Future of Meat and Dairy

Consider how modern science can make livestock more amenable to the local food system and global climate struggle. It is possible, even essential, to marry old-fashion techniques with cutting-edge research to create healthier animals who produce a more nutritious product while lessening their impact on the environment. For example, cows may be given a novel methane inhibitor as a supplement in their food. Studies show this will cause them to produce dramatically less methane from their flatulence, helping both regional air quality and also global warming, with the added bonus of allowing cattle to gain weight, and therefore value, easier and more quickly.[22] Cattle can also

---

22 Jeff Mullhollem, "Feed Supplement for Dairy Cows Cuts Their Methane Emmissions by About a Quarter," Penn State Department of Agricultural Science, accessed April 18, 2020.

be selectively bred to consume less feed and water while growing faster so they may be slaughtered sooner. Shortening cattle lifespans and reducing their resource consumption means lessening their environmental impact even more. The combination of these advances along with time-honored land management practices will make ruminant livestock better neighbors and let small ranchers keep larger herds and flocks on healthier land. This will make CAFOs not just more obviously repugnant, but also economically obsolete. Ultimately, regions will have stronger networks of small-scale meat and dairy producers on which their nearby consumers can rely for high-quality, healthier, and sustainable animal products.

Thus, an amalgamation of Savory's method of holistic land management and other similar ecologically-sound practices, a fundamental transformation of the American meat industry, the home-raising of small livestock, and using modern science to breed and raise better livestock will be crucial to helping bring meat production back down to earth. But the hard truth is we will have to learn to make do with more expensive meat and dairy and to eat less of it. We have a very long way to go before our meat industry will make a positive impact on our environment and on our health, but empowering local regions to become better stewards of their pastures while supporting small, independent producers is a vital first step in that positive direction.

CHAPTER FIVE

WILD SALMON AND THE LIMITS OF LOCALISM

In a world often plagued with reports of environmental catastrophe, Alaska is celebrating a spot of good news: the Last Frontier's fisheries are doing better than ever. The early to mid-1900s saw a precipitous decline in Alaskan salmon stocks due to overfishing caused by innovations in fishing and cannery technologies. In 1953, President Dwight D. Eisenhower declared Alaska a federal disaster area due to the incredibly diminished salmon catch. However, since statehood was declared in 1959, the management of the fisheries devolved from federal administration to the newly formed Alaska Department of Fish and Game. In 1973, fisheries created a limited entry permitting system to ensure escapement of adult spawning salmon to lay their eggs and renew stocks. The Alaskan salmon population, as a whole, has rebounded dramatically and continues to provide sustainable high yields year after year.

However, this story is not entirely a happy ending. If Alaskan salmon fisheries are doing so well, why are the smaller fishermen doing so poorly? It's no secret in the state's coastal communities that times have only been getting tougher—the good seasons aren't as good as they used to be, the bad seasons are even worse, and profit margins get thinner and thinner every year. A cursory look at the evolution of the Alaskan seafood industry shows why—since the 1980s, North American fisheries saw increased competition from foreign markets, especially those in East Asia. Low cost producers of shrimp in Thailand and fish farms in China continue to undercut the prices of wild salmon producers in North America and processing operations in China have become increasingly inexpensive.

To remove the pinbones from salmon fillets, the labor cost in the US would be one dollar per pound compared to just twenty cents in China. Nowadays, China is Alaska's largest export partner in this sector, taking in 35 percent of salmon by tonnage in 2015.[23] This has led to Alaskan salmon being no more expensive to consumers today than it was in the 1980s when adjusted for inflation. However, fixed costs have grown across the board from fishing licenses, to fleet maintenance, to labor and supplies costs, and more.

While Alaska offers some loans and grants to help small-scale fishing operations meet those tremendous operations costs, the sums are often meager and are difficult to get without already having a certain number of hours on the water for your fleet. This makes the industry very difficult to enter for small, new fishing businesses. Thus, the large

---

23 *Alaska Seafood Export Market Analysis* (Juneau, Alaska Seafood Marketing Institute, 2016).

players use the tremendous amounts of capital they've already invested to continue to collect more and more fishing permits for an ever-expanding fleet, pushing out smaller players. This situation is even worse when fishermen wish to employ sustainable techniques like hook-and-line fishing. Hook-and-line does result in dramatically smaller catches than fishing with nets. However, the alternative, trawlers dragging huge nets along the ocean floor, obliterates ocean floor ecosystems and does not allow for most fish to escape capture, swim upstream, spawn, and thereby provide for next year's harvest.

Further complicating this mess are decades of failed neoliberal free trade policies designed without the interests of small fishermen in mind. Alaskan salmon fishermen have always had to deal with changing prices canneries and distributors are willing to pay for their catch each year. This is usually due, in part, to demand factors such as how much salmon consumers are buying and how much they're willing to pay for it. Also consequential are supply factors like how big the salmon catch was and how big the catches of substitutes, like swordfish or tuna, were that year. However, since being opened up to international competition, these price fluctuations have only become more and more volatile. One good year may net a small fishing operation one hundred thousand dollars in profit only to be hit hard with less than half the next year. This unpredictability makes it increasingly difficult to plan for the future and weather the bad years unless you are a larger operation with sufficient liquid assets to cover any unexpected losses.

A Market-Based Solution for Salmon Fishermen

So what's an Alaskan fisherman to do? Environmentalist-turned-professor Nicholaas Mink asked himself the same question when he was working with the Sitka Conservation Society in Sitka, home of Alaska's largest harbor system and sixth largest American port by value of seafood harvest. While he was building educational programs to protect Alaskan small boat fisheries, he realized top-down policy-based solutions could only do so much. After returning to academia to teach food systems at Knox College, he grappled with how to best develop an alternative market-based model that could incorporate sustainable practices while giving fishermen higher, more predictable incomes.

In 2012, he launched Sitka Salmon Shares with a new paradigm: better for fishing families, fresher for domestic consumers, and minimal environmental impact. Fishermen are more insulated from the swings of the global commodities market because Sitka Salmon Shares promises them a "price floor" beyond which the price they will buy their salmon will not fall. This is because the company not only owns the fleet, but also the packaging facility and the distribution hub. This ownership of the supply chain offers them greater flexibility in terms of pricing and helps to keep costs down. These savings are passed directly on to their fishermen, allowing them to sell their fish to Sitka Salmon Shares for an average of 18 percent more per pound than their large-scale counterparts. In some particularly volatile years, this can spike to as high as 40 percent more as global prices plummet and the price floor kicks in.

Nicholaas Mink, Founder of Sitka Salmon Shares, handling fish in their Sitka facility

The benefits of this model extend far beyond purely economic stability. Fishermen get to play a large part in corporate governance. Four rotating seats on the board are reserved for fishermen and they also get a share in the company's equity, the value of which has only been growing. On the environmental side, their fleet exclusively employs hook-and-line fishing instead of nets to promote healthier salmon stocks. Also, fish are shipped across the United States using only boats and trucks. Planes, although theoretically faster, use roughly forty times more fossil fuels in their operation than land and sea transportation options. Plus, the fish are all wild-caught and from a well-managed fishery as opposed to environmentally horrendous farmed fish. As for the consumer, Americans who sign up for the monthly "share" get seafood delivered to their house which is weeks fresher than anything they can find in their local supermarkets due to the company's collapsed supply chain.

This presents a glimpse into how to create a conception of food sovereignty that departs from more rigid definitions of food independence. Those of us who advocate for producing as much food locally as possible understand it may not always be economically viable or environmentally sound to do so—and that's okay.

For example, Southern California is a notoriously hot and dry place that is not naturally suitable for large-scale cattle farming. It can certainly be done, and beef cattle are being raised in Southern California to this day, but the amount of grain that has to be trucked in and the water and electricity used to maintain the grass in the cattle's pasture is dramatically higher than in cooler climates. Therefore, a grocery store owner in Los Angeles who wants to buy beef at wholesale quantities while being cognizant of environmental impact and price may look to places further away.

British Columbia is considerably farther away from Los Angeles than a ranch in Bakersfield, but cattle there can enjoy plenty of cheap, plentiful local grain and grass that grows without intensive irrigation. Seeing as British Columbia has a comparative advantage in beef production, it makes sense both fiscally and environmentally for their ranches to supply stores in Southern California than it does for Southern California to create the beef itself. In the same vein, it would be prohibitively expensive and environmentally catastrophic for a city like Phoenix, Arizona to attempt to produce its own local salmon on a fish farm. Phoenix can help Sitka, which cannot consume all the fish it produces, maintain its own food sovereignty through fair trade practices which honor the work small-scale fishermen put into their harvest. In return, Phoenix gets access to a reliable, fresh, and sustainable supply of wild-caught salmon instead of having to compete in an area in which it has no advantage.

Clearly, food sovereignty is not about complete food independence for each and every city and its hinterland. Rather, it is about communities having ownership over their food supply, whether they are producing it for their own consumption or for another community's needs. While localism is indeed a major part of the movement, recognizing the limits of local food production in our modern globalized and urbanized world is essential for crafting the realistic and resilient food systems of tomorrow.

Wild Foods Are Everywhere

In addition, an often disregarded but increasingly visible aspect of burgeoning local food systems is the wild foods movement. For many urban American consumers, the only wild food they may come in regular contact with is maple syrup or wild-caught fish and shellfish, such as the aforementioned Alaskan salmon. In rural areas, this list may include some wild berries and nuts, roadkill, or hunted meats like venison or elk. However, wild foods are popping up everywhere from high-end restaurants to local farmers markets. More and more Americans turn to their own backyards and find common "weeds" like dandelions and wild onions are just as good (if not better and fresher) than anything they can find in a supermarket.

Foraging, or the act of going into a wild space to collect edible plants, fruits, seeds, nuts, fungi, leaves, herbs, and insects, is a method of resource extraction that is as old as humankind itself. However, with many humans now living in cities and suburbs with sporadic contact with the natural world along with the convenience of the supermarket, convenience store, and local restaurant, foraging has been thrown by the wayside and seen as primitive or even dangerous. Foraging may also be seen as something low-class and therefore give you an undesirable reputation amongst your neighbors. A prime example of this is the American pokeweed, a wild plant found all over the United States. This plant, while toxic if eaten raw, produces leaves that can be made edible through a painstaking, boiling process and was a common food in the Southern United States. However, picking young pokeweed leaves in the Spring became associated with low-class laborers and thus, became known as a "poor man's food." So the process largely died off during the twentieth century simply due to the social stigma.

Another barrier to the proliferation of foraging is the knowledge gap. Most Americans are not raised with a deep understanding of their local fauna and flora. While all cultures around the world have these connections to their local environs, they too are oftentimes forgotten or not passed down by immigrant families during the assimilation process. The countless culinary and medicinal applications of endemic plants is not taught in schools. These skills are thought to not be applicable to our modern society and have been discarded in favor of the traditional school subjects. Of course, it's hard to blame our teachers for this when they themselves wouldn't know where to begin when it comes to feeding one's self from one's own backyard. The sad truth is the vast majority of the knowledge indigenous people had about the ecosystems of North America has been lost due to the violence of colonialism and the hegemony of European epistemologies. The suppression of autochthonous food knowledge and traditional foodways was a vital part of the colonial experiment in divorcing native people from their ancestral lands, instead creating a culture of dependency on the mother country for their needs. While the era of colonial expansion in the Americas has largely ended, its ramifications on not just what we know about our own landscapes, but how we can get to know them better, are profound and long-lasting.

Nevertheless, there are folks in every state using their local ecosystem to feed themselves and others. A prime example of this burgeoning movement is Sam Thayer, one of the Midwest's most prolific wild foods experts. Sam has been foraging the woods of his native Wisconsin for decades, travels around the United States conducting foraging workshops, and has authored three books on wild edibles. He was even inducted into the National Wild Foods Hall of Fame (yes, this actually exists) in 2002. While he lives mostly off the grid, he also runs a small business, aptly named Forager's Harvest, selling the foods he forages wholesale, such as wild rice, maple and birch syrup, apples, hickory and acorn oil, and a variety of native berries, just to name a few. This cottage industry in the little town of Bruce is growing and now more than half of his family's food comes from foraged components. Sam is a passionate advocate for wild and foraged foods and is hopeful that as more people are exposed to them, they will start to open their hearts and pantries. His vision is simple: If people can take just 10 percent of their food consumption and replace that with things they harvest from the wild, he reckons there is an all-around benefit to the consumer, to the economy, and to the ecology.

Sam Thayer giving a foraging workshop in the woods of Wisconsin

So how do we begin to reintegrate wild foods into our food system? Sam's homesteading is a good departure point, but that lifestyle is out of reach for the vast majority of Americans in urban and suburban areas. Getting Americans to reconnect with their wildernesses as a source of food is a complex problem, as Sam Thayer explains:

"There's an underlying assumption that the appropriate way to integrate is through the market economy, but that can't be the starting point, and the reason it can't be the starting point is because in order for something to enter the market economy,

there must be a market. There must also be the knowledge and the skill of producing the marketable item, and we have neither thing."

Clearly, integrating wild foods into our diets must start at a more grassroots level. It has to begin with ordinary people reading a book or searching the internet to find some readily accessible foods in their own backyards. When even just a few people begin to confidently source just a fraction of their diets from their environments, it will expose more and more people to a lifestyle they never considered. Joining a local organization dedicated to foraging and wild foods workshops is also a great way for individuals to feel empowered to incorporate foraged foods into their plates. Wild foods are some of the most nutritious, tasty, and pesticide-free foods you can find—and they're free!

Of course, this is not to say one cannot make a decent living off foraged foods. For example, Hammon's Products Company, based out of Stockton, Missouri, has been mass-producing black walnuts since 1946. The woodlands of the Midwest produce an abundance of black walnuts, most of which would go uneaten. Hammon's has no orchards and instead relies on ordinary people going out into the woods, picking fallen walnuts off the ground for their product, and selling them at one of their 215 buying stations across eleven states. This business model has produced an abundance of cheap black walnuts that can be sold across the United States at stores like Publix, Walmart, Costco, and more.

Beyond the health, economic, and flavor benefits (of which there are many), there is also something profoundly sacred about the foraging experience. To feed one's self using just your knowledge, your hands, and the natural world around you is perhaps the most satisfying food preparation experience one can have. Going to the supermarket is just no comparison. It's one of the reasons Sam Thayer is so passionate about his work, as he explains:

"The most important thing that forging can offer immediately to our society is that it is impossible to forage for food without experiencing gratitude. I like to remind people that gratitude is an instinct just as much as fear, just as much as jealousy... and it's there for a reason. Our ancestors could not have survived without gratitude."

Perhaps if more Americans decided to learn a bit about the plants around them and see if they can incorporate just a bit of their environment into their diets, we would all develop a greater respect and love for the landscapes in which we live. Maybe this will give our country a new perspective and heightened appreciation for the natural world and give us the desire and willpower to protect it.

All in all, wild foods have been overlooked and under-utilized for far too long. It is incumbent on all of us to learn about the ecologies in which we carry out our everyday lives and maybe even contribute just a little bit to rediscovering the library's worth of knowledge that has been lost. As a case study, Sitka Salmon Shares offers an integrated new model for how small fishermen can be better served by organizations that share their values, care about the environment, and exist to serve their economic well-being as opposed to lining the pockets of cannery owners or executives of seafood distributors. Returning the sovereignty over Alaskan fisheries to their communities means more than just devolving regulatory control to a state or local level. One of the most important facets of food sovereignty is improving the lives of the people who grow, fish, harvest, process, distribute, cook, and sell us what we eat. Sitka Salmon Shares has put the ownership and economic power back in the hands of these small family fishing operations while sheltering them from the worst of what the globalized food system has wrought on their industry. This model is easily replicable in other communities throughout Alaska, North America, and even the world.

PART TWO

TRANSFORMING THE FOOD ECONOMY

CHAPTER SIX

RESTAURANTS AT THE HUMAN SCALE

In late October of 2012, Hurricane Sandy barreled into the heart of the Mid-Atlantic seaboard. Gusts of nearly one hundred miles per hour ravaged the Northeastern United States, causing tens of billions of dollars in damages and resulting in the deaths of 117 people. While making landfall in Southern New Jersey, the hurricane still had a deadly effect on New York City. Widespread flooding, six-alarm fires, and an electrical blackout for over eight hundred thousand New Yorkers spelled disaster for the local economy. Many restaurants were hit hard: the blackout meant dining rooms and kitchens couldn't be illuminated, refrigerators and appliances couldn't function, and many employees couldn't get to work.

A Restaurant versus A Hurricane

While eateries across the city closed their doors, one little spot in the East Village was serving up hot meals all day long. With nothing but headlamps to illuminate the gas range and a dining room filled with candles, Chef Will Horowitz of Ducks Eatery and his team were pounding out mouth-watering dishes to hungry city-dwellers around the clock. During the day, regulars were treated to classic red beans and rice with whatever protein could be donated from other restaurants whose freezers were rapidly warming. Nighttime arrived and the kitchen flipped to serve up fresh grilled Alaskan snow crab with whipped lardo for the city's homeless who were pushed out of flooded shelters. In the end, Ducks Eatery didn't miss a single day of business throughout the entire disaster and they won the respect of local denizens who still pull up a stool to their counter every week.

So what's the secret? How could this scrappy six-hundred-square-foot Lower Manhattan restaurant make it through the storm while larger, more famous establishments couldn't handle the pressure? The recipe to Ducks' success lies in the very DNA of the restaurant: While other restaurants had many of their ingredients rapidly turning without refrigeration, Ducks Eatery prided itself on integrating heritage techniques of food preservation, such as brining, pickling, and smoking, to create food that was not just delicious, but also able to be shelf stable for months without the need for freezing. Moreover, Ducks Eatery sources their fresh ingredients from local producers, meaning while distributors' supply lines were disrupted due to fallout from the hurricane across the region, they were able to get by on what was already in the city. Plus, it certainly helped that Will lived right above the restaurant, a choice made so he could smell when items in the smoker were done through the vents.

The phenomenon of the great American chain restaurant is something familiar not only to Americans, but to the countless nations around the world to which they've been exported. It seems no matter where you go, you are bound to find the familiar golden arches, crispy chicken sandwich, or fat burrito expertly wrapped just like you can get down the street from your home. Uniformity may be the single greatest achievement of our globalized capitalist economy. But did we lose something along the way?

Fast Facts on Fast Food

Corporate fast food, from french fries to fajitas, is good for an inexpensive, dependable meal, but we all know the real costs of that food are all externalized. Fast food is notorious for detrimental health effects popularized in documentaries like *Super Size Me*. Seeing over one-third of adults consume fast food on a given day, it is no secret the industry contributes heavily to the epidemic of obesity and diabetes in the United States and to the countries it is exported. The Center for Disease Control reports over the past two decades, the percentage of adult Americans who are obese increased from 30.5 percent to 42.4 percent and the prevalence of severe obesity increased from 4.7 percent to 9.2 percent.[24] This meteoric rise has been increasingly linked to a variety of deadly diseases such as heart disease, cancer, and more and is a direct result of the proliferation of fast food in our diets over the past fifty years.

Environmentally, the massive global supply chains needed to sustain global fast food empires cause catastrophic harm to

---

24 Craig M. Hales, M.D, et al. Center for Disease Control, National Center for Health Statistics, *Prevalence of Obesity and Severe Obesity Among Adults: United States, 2017–2018* (Hyattsville, Maryland, 2020).

the planet. From beef cattle raised on deforested Amazonian land to make the burgers, to the huge water-wasting, monoculture plantations used to grow the produce, to the untold fuel costs to ship it all around the world, fast food is one of the dirtiest ways to provide the planet with meals. Moreover, food service is the most energy-intensive commercial activity in our modern economy and is the top user of refrigeration technology, which often employs gaseous coolants which are harmful to the atmosphere and ozone.[25] Clearly, fast food is an industry of which the earth could do with much less.

Furthermore, economic effects of piping dollars out of local hands and into the pockets of corporations and their shareholders holds economic benefit. Fast food companies infamously pay their employees as little as possible with few to no benefits while lavishing their CEOs with incredible wealth. Former McDonald's CEO Steve Easterbrook took home $21.8 million in 2017 while tens of thousands of employees relied on food stamps to feed themselves. In Ohio, McDonald's employees constitute the second largest group of employed individuals on food stamps behind Walmart employees with nearly ten thousand employees accepting food aid. So while American tax dollars subsidize bottom-of-the-barrel wages for their neighbors, profits flow upward out of communities and the working people who form them and into the hands of Wall Street and MBAs at corporate offices.

It's no wonder the food is cheap—everyone is paying!

---

25 Terry Davies and David M. Konisky, *Environmental Implications of the Foodservice and Food Retail Industries* (Washington, Resources for the Future, 2000).

The Farm-to-Table Connection

Ducks Eatery represents a new paradigm in modern dining that isn't very new at all. While something of an outlier on the busy streets of Manhattan, Ducks shows there's a space for restaurants that embrace time-honored methods of food production and are truly *of* a place and not merely in it. Across the country, chefs are flipping the script and opening dining rooms at the human scale, working with their producers and surroundings to craft a better dining experience for their guests, workers, region, and planet.

Running a restaurant at the human scale lends itself more easily to building personal relationships between the restauranteur and the suppliers on which they depend. A classic story of a cook working closely with his producer is playing out in a small bakery in the middle of our nation's capital. Jonathan Bethony is a baker in the trendy DC neighborhood of Shaw in the heart of downtown Washington. While young, he has made quite a name for himself in the baking world with his Pain au Levain winning the top prize in the Wholewheat and Wholemeal category at the 2019 Tiptree World Bread Awards. However, Jonathan is much more concerned with fostering a deep connection to the land and the local organic farmers who work it. In his own words, his business is much more than just a high-quality bakery:

"Our philosophy in how we run the bakery comes from how we source for the bakery and how we engage our community...Following the guiding principles of regenerative agriculture, which ties in to something deeper. For lack of a better way of saying it, it is the healing arts... about providing a staple, but with the

intention to make things better for the next generation. Those threads go all the way through [our work] from start to finish."

To accomplish this weighty goal, he maintains close working relationships with local farmers, letting his menu be dictated by the needs of the fields that bear his ingredients. A classic example of this is his "horse bread," a loaf made from a hodge-podge of millet, mustard seed, field peas, and more. The inspiration for this creation was in the land: it was the blend of seeds his partner farmer had sewn to rejuvenate his land and keep it primed for future plantings of traditional wheat varieties Jonathan also uses. In another case, the bakery was interested in sourcing its ingredients for its tahini cookies to a local supplier. So Jonathan asked one of his partner farmers if they wouldn't mind growing benne seeds for the bakery, to which they agreed. This symbiotic relationship between kitchen and field promotes a more sustainable and climate-resilient product and a wholesome interdependence between a city and its hinterland, catalyzing local economics instead of draining them. Farmers unlock the benefits of having a guaranteed buyer for their crops or a reliable, flexible partner to call up when other contracts fall through and they have to unload excess supply. Restaurants gain a dependable supplier whose quality they can trust and thereby forge a more resilient supply chain for their business. Also, in a grander sense, it bolsters a region's control over its own food supply while fostering a greater understanding and appreciation of the rhythms of the local land and seasons on the part of the business owner and their patrons.

Just like Seylou, upon cultivating these meaningful connections with the purveyors of their ingredients, Ducks

Eatery is also able to more easily identify opportunities most other restaurants would miss. For example, by having conversations with their local butcher, Chef Will Horowitz is able to find great cuts of meat that were typically being ground up and sold to dog food companies due to low demand. Products like goat feet, which may usually be too gelatinous or unfamiliar for the tastes of most Americans, became instant hits at the restaurant when smoked and creatively prepared.

Ducks Eatery gets a cheaper cut of meat from which they create a high-quality product, creating profit and reducing waste, while the butcher gets paid more than under his original plan to sell to the dog food manufacturer. This creates a win-win situation that would not have been possible without the human-scale connection between two local small business owners. This kind of commerce, when magnified across a region, can foster a great interconnectedness between enterprises in a community, adding tremendous value to the local economy and strengthening the resilience and control of the local food system.

Revolutionizing American Gastronomy

Another positive lesson from the aforementioned goat feet is the broadening of the American palate. It follows if we find a greater variety of things we are willing to consume from our local food system, less food will be wasted. This raises the overall value of food in local economies, so more money can go into the pockets of food producers. Plus, consumers can enjoy a greater variety of vitamins, minerals, and flavors. Restaurants have a critical role to play in expanding our culinary landscape by creating approachable, artful meals with novel ingredients.

A great example of this is the dramatic transformation of sushi in the United States. Before the 1970s, you'd be hard-pressed to find many sushi fans outside of the Japanese-American community. Indeed, the thought of consuming raw fish was extremely foreign and unappealing to many American diners. But with a new wave of Japanese immigration in the 1970s came a renewed interest in Japanese culture and more and more people came to love the familiar selection of seaweed rolls filled with rice and seafood. Also helpful was American chefs both with and without Japanese ancestry combining traditional elements of Japanese cooking with American ingredients and preferences, yielding such sushi bar standards as the California roll. Nowadays, sushi is as familiar to the American culinary landscape as pizza and tacos. Even in major supermarkets, you can typically find sushi rice, nori, sushi rolling mats, and other accoutrement with relative ease. This same concept can be applied to local ingredients that are not yet commonly embraced by the American gastronomical lexicon.

Restaurants have again been at the forefront of developing the American palate by utilizing locally produced meats like bone marrow, beef tongue, chicken hearts, and typically foraged produce like pawpaws, ramps, and morel mushrooms. These foods often have storied histories as a part of the diets of various American Indian cultures as well as in the cuisines of cultures all over the world. Typically excluded from our massive food supply chains due to limited popularity or structural issues (such as pawpaws, which are harvested soft and therefore do not travel well on cross-country trucks), chefs are able to breathe new life into heritage ingredients by sourcing them locally from farmers and foragers alike. By working with their region's bounty and using more of what is locally available to them, restaurants are consistently pushing the needle on the evolution of American cuisine and are in the process of making regions more autonomous, sustainable, prosperous, and secure in their food production.

National Chain, Local Impact
However, it is naïve to think our culinary landscape will be ruled by mom-and-pop eateries churning out locally sourced, innovative dishes at every turn. There will always be a place for corporate dining in America, so it's valuable to seek out examples of major restaurant chains who are making positive contributions to the communities they serve. An exemplary case can be found in the chain who has revolutionized the fast-casual dining industry: Sweetgreen. The salad-slinging spots with their chic wood-and-concrete interiors have become a staple for health-conscious eaters in major cities across the United States.

From Sweetgreen's inception, the company has maintained a serious commitment to cultivating a positive impact on their locations' local economies and environments. Menus change with the seasons and are dictated by what local farmers are planting. Ingredients for each location are sourced as locally as possible from farmers who are practicing sustainable agricultural techniques. In fact, every store features a chalkboard where you can find the source farm and its location for every item on the menu. Everything in the store, from the cups and bowls to the forks and napkins, are compostable and compost bins are provided in-store. On average, the company is able to divert 60 percent of waste from landfills and compost 75 percent of all food scraps. By regionalizing their supply chains and taking stewardship of their local environments, Sweetgreen is able to run a national corporation at the human scale.

Restaurants have a clear role to play in the decentralization of the American food system. They are on the frontlines of food trends and have tremendous influence on the American economy and popular culture. In 2018, the industry realized $833 billion in sales and employed over 15.1 million people in the United States alone.[26] Imagine what a sea change it would be if those restaurants utilized their massive, combined buying power and transitioned just a quarter of their supply chain to local sources. More food distributers would take notice and transition their operations along decentralized lines as well. This all starts with consumers voting with their dollars and making it a point to make more room in their dining budgets for restaurants making use of the bounty of their regions. This will make restaurant supply chains more resilient, allow city and hinterland economies the opportunity to forge deeper and more profitable connections, and reduce the carbon footprints of restaurants and food distributors. In this way, restaurants have the potential to lead the way in realigning our food economy to better the communities they call home.

---

26 Hudson Riehle, et al., *Restaurant Industry 2030: Actionable Insights for the Future* (Washington, National Restaurant Association, 2019).

CHAPTER SEVEN

HOW THE (VEGGIE) SAUSAGE GETS MADE: ADDRESSING FOOD WASTE FROM ALL SIDES

Paula Schafer smiled as she looked through the fence at the meatpacking plant that long dominated the skyline of her home in Montgomery County, Pennsylvania. When she made feeding the hungry in her community a full-time job, she hadn't imagined how difficult some folks could be about donating leftover foods. Just a week prior, all that stood between her and delivering tens of thousands of pounds of beef to local food banks who were struggling under the pressure from the Coronavirus was logistics: The plant's loading dock was incompatible with her small, refrigerated van, plant management said, and therefore they would not be able to accommodate her request. But Paula was not one to give up—she continued to appeal to management until finally, they relented. As she and plant workers hauled the frozen meat into her van, she could only reflect on the countless meals she'd be able to provide and the countless bellies that will be filled as a result, all because she didn't give up. This is the work behind Montgomery County Anti-Hunger Network, a nonprofit that helps local food banks acquire and distribute food more efficiently, getting food to where it's needed more effectively. Paula and her driver are the only employees and they are committed to ending hunger nationwide by starting right where they live.

Paula Schafer is the Executive Director of MontCo Anti-Hunger Network

Food Waste: A Global Problem

One of the most dramatic yet hidden problems with our current food system is the issue of waste. Estimates show the percentage of food wasted in the United States and in all countries around the world is between 30 to 50 percent.[27] This means that roughly 1.4 billion hectares of farmland globally is used to produce food that never gets eaten. This horrific environmental track record is aggravated by all the inputs wasted farmland uses: food waste contributes to about 4.4 billion tons of carbon dioxide alone every year (not counting all the methane released by rotting foods in landfills) and uses 250 km^3 of water each year, which is more than half of the water in all of Lake Erie.[28] [29]

In the United States alone, where more than thirty-seven million people struggle with hunger, it's likely more than one in three pounds of food ends up in the trash. Such a staggering statistic becomes even more upsetting when considering that out of those thirty-seven million people, *12.5 million of them are under 18,* making children the most food-insecure demographic in the United States. Both of these numbers are likely to skyrocket to over fifty million and eighteen million respectively due to the economic effects of the coronavirus pandemic.[30] The problem for young people doesn't end when they go off to college either: 39 percent of undergraduate

---

27 George Aggidis FIMechE, et al., *Global Food: Waste Not, Want Not* (London: Institution of Mechanical Engineers, 2013).

28 Emilie Wieben, *Save Food For a Better Climate: Converting the Food Loss and Waste Challenge into Climate Action* (Rome, Food and Agriculture Organization of the United Nations, 2017).

29 Pietro Bartoleschi, *Food Wastage Footprints* (Rome, Food and Agriculture Organization of the United Nations, 2013).

30 *The Impact of the Coronavirus on Food Insecurity* (Chicago, Feeding America, 2020).

students have experienced food insecurity, a need traditional food banks have struggled to address. This has led to the proliferation of campus food pantries throughout the country as local pantries in these students' adopted towns have scrambled to deal with this annual surge in need.[31]

In an era of history with unprecedented food production, the thought of so much food going to waste instead of to hungry mouths is shameful. While a lot of welfare programs targeting food access, such as the Supplemental Nutrition Assistance Program (SNAP, also known as food stamps) or National School Lunch Program (NSLP), are funded and directed at the national level, the federal government has been largely absent with action on the food waste issue. Exploring how communities can reduce food waste is essential for helping regions build more sustainable and inclusive food systems.

When approaching this issue, the first thing to consider are the main sources of all that food waste. While the global rate of food waste is similar across all countries, the sources of the most food waste diverge based on development in their location along the supply chains. Developing countries tend to experience most of their food waste during two main phases: production and transportation. During food production, crops may be destroyed by natural forces like droughts or storms, leading to losses for both the farmer and the food system. This waste is somewhat unavoidable in

---

31 Christine Baker-Smith, et al., *#RealCollege 2020: Five Years of Evidence on Campus Basic Needs Insecurity* (Philadelphia, The Hope Center for College, Community, and Justice at Temple University, 2020).

poorer countries, but developed nations with drought-resistant crops, modern irrigation systems, and crop insurance suffer much less in this area.

Also, and perhaps most infuriating, is the issue of "culling," in which farmers discard perfectly good produce based purely on aesthetic reasons. This is because many supermarkets will only accept carrots that are perfectly orange and straight, bananas that are bright yellow and curved, tomatoes that are familiarly rotund, and so on. In each harvest, as much as one in seven trucks of produce can be rejected by retailers based on size, color, or shape depending on how strict the producer's (or more likely the distributor to which the producer sells his crop's) standards.[32] Jeremy Kranowitz, the CEO and President of Keep Indianapolis Beautiful, Inc., a nonprofit organization dedicated to preserving the natural spaces of the city of Indianapolis, offers a poignant example:

"Green beans consumed in the UK are frequently grown in Sub-Saharan African countries, where it's not a native crop and it's not something that the people in those African nations consume regularly or know what to do with... [In the UK,} they were selling the beans in 10-centimeter lengths because that's what fit in the 10-centimeter-wide bin in the grocery store. But the beans didn't know that they had to grow to be exactly 10 centimeters. So, they would pre-trim them and snip the ends off them get them to be the right length, but then that significantly hastened the speed at which they decomposed. Also, if

---

32 Dana Gunders, *Wasted: How America Is Losing Up to 40 Percent of Its Food from Farm to Fork to Landfill* (New York City, National Resource Defense Council, 2012).

you grew a 12 centimeter bean, they weren't doing anything with the two centimeters they were snipping off—it was just going to waste [and this was being done] in countries where there is great food insecurity. So, all kinds of problems [were occurring] all because the bin in the grocery store was 10 centimeters wide."

While the problem seems absurd, it is precisely this sort of corporate idiocy that plagues food systems in both developed and undeveloped nations. Furthermore, transporting perishable foods in underdeveloped market sheds even more stock due to lack of refrigeration technology and inefficient supply chain logistics. What may be fresh for two weeks in a refrigerator may only be left at ambient temperature for a few days. However, the vast majority of the food that reaches the consumer is consumed and relatively little is wasted in the home.

On the other hand, countries with integrated, high-tech supply chains have more efficient harvests (although they still oftentimes practice culling) and can get more of their produce to consumers faster and fresher. But Western retailers and consumers waste a lot more of the food they purchase than their counterparts in other parts of the world. American farms and businesses discard over six billion pounds of food based on its appearance every year.[33] That's equivalent to the weight of *over 16,400 Empire State Buildings!* Plus, expiration dates on food are often inconsistent between brands or are

---

33 Dana Gunders, *Wasted: How America Is Losing Up to 40 Percent of Its Food from Farm to Fork to Landfill* (New York City, National Resource Defense Council, 2012).

just ambiguous. Many retailers will deal with food that is past its "best before," "use by," or "sell by" dates by simply throwing it in the trash, regardless of whether or not the food is actually still good. This is a problem that also affects consumers, who may be led to believe perfectly good food is unsafe.

For example, a package of raw chicken thighs may have a sell-by date of April 26, but if kept frozen, it may still be perfectly good to eat through the end of the year. Packaging also contributes to food waste in a major way: Some packaging may be too large, allowing for some food to spoil as the portions are too large to be consumed. Some may make it hard for all the product to be used due to structural issues, such as honey stuck to the sides of a container. Some foods are sold with improper packaging, causing them to spoil sooner, such as raw meat sold in parchment paper rather than an airtight plastic wrap. Confusion over expiration dates, aesthetic qualities of food, and packing issues all add up. According to the Food and Agricultural Organization (FAO), the average North American and European consumer wastes between 210 to 250 pounds of food each year as opposed to between thirteen to twenty-four pounds by the average person in Sub-Saharan Africa.[34] The comparison is dramatic, to say the least. To alter the course of food waste in the United States, those organizations which are closest to consumers, such as local governments and nonprofits, have the biggest role to play.

---

34 Dana Gunders and Jonathan Bloom, *Wasted: How America Is Losing Up to 40 Percent of Its Food from Farm to Fork to Landfill* (New York City, National Resource Defense Council, 2017).

Fighting Food Waste in America

In our country today, the most important way we can combat food waste at the local level is recovering good food that would normally be discarded. We have a well-known, last-mile problem with prepared food waste in this country. Plenty of people are familiar with the problem of bakeries discarding hundreds of pounds of perfectly good pastries at day's end so they can offer freshly-baked products to their customers each day. On a larger scale, establishments like stadium restaurants may whip up hundreds of piping-hot pizzas only to trash half of them, uneaten, after the game is through. It may seem nonsensical, but this is in part due to government regulations requiring donated prepared foods to be well-labeled with ingredients and allergen warnings, putting too great an onus on small businesses that may not have the time or resources for such a task. There is also the issue of individuals or businesses being afraid their donated food may end up getting someone sick by some small chance. What many of those concerned may not know is their liability is covered at the federal level by the Bill Emerson Good Samaritan Food Donation Act, which protects well-intentioned folks who donate food that appears safe from legal harm. Even better, there has never been a single incidence of the act being invoked, meaning this fear is fortunately unfounded.

However, this is not a problem faced by many supermarkets as their food comes pre-packaged and they have a general counsel who understands the ins and outs of food law. A region's food banks, however, which are institutions who store stocks of food to provide to people in need, are typically run by a potpourri of different organizations of various sizes and capacities which oftentimes may not communicate and have trouble soliciting donations from larger food distributors and supermarkets. This is especially true for smaller food pantries which may not have the capabilities to handle large, regular food shipments from a supermarket which could quickly overwhelm their inventory. This is a lose-lose-lose scenario: The company misses out on a philanthropic tax write-off, the food bank misses out on an opportunity to stock more wholesome food, and most importantly, the beneficiary misses out on a meal.

So what can be done to lessen food waste and put what's left to good use? This is where something like Paula's organization comes in: The Montgomery County Anti-Hunger Network can leverage its participating food banks to send the incoming food donations exactly where they need to go to have the most impact. This results in fewer donations being refused, less food wasted, and more meals provided. Ms. Schafer calculates that her organization saved about 121 tons of food in just 2019 alone. That's nearly fifty thousand meals over the course of a year and that's just in one suburban county in Pennsylvania! Imagine how many people could be fed if similar initiatives were undertaken in every county around the country. Local initiatives like this are critical to meeting the challenge of the last-mile problem in food waste distribution. Indeed, organizations operating at the local and hyperlocal level may be the only ones which are enmeshed in the fabric of the community and nimble enough to understand and adapt to the dynamic landscape of a region's food scene.

Rescuing food that does not typically make it to shelves is also important. Even better is when those market inefficiencies become market solutions. Such a solution is realized in

Misfit Foods, a Brooklyn-based startup born out of George-town University with a mission to save "ugly produce." Initially, founders Philip Wong and Ann Yang started their company as a way to turn fruits and vegetables that didn't fit supermarket aesthetic requirements into healthful and tasty juices. Since, the company has pivoted to a more niche, less crowded market: creating sausages made of 50 percent humanely-raised chicken and 50 percent fruits and vegetables that would otherwise have gone to waste. There have also been a variety of produce delivery startups like Hungry Harvest and Imperfect Foods which will deliver regionally sourced ugly produce right to your door for prices well below traditional supermarket rates. Through entrepreneurship, innovators are helping producers cut down on their food waste and make some extra cash while they're at it.

On the consumer side, Local governments have also stepped up to the plate by introducing citywide composting programs. Boise, the capital of the State of Idaho, for example, has a robust "Curb It" program complete with curbside pickup for house-hold food waste and two locations where participating citizens can go pick up free compost for use in their home gardens. Citizens are incentivized to participate with free, year-round leaf collection and a free compost cart to be placed out with their trash for collection. Smart policy like this creates a local cycle of good choices which can only end up with positive externalities in the long run, such as greener streets, cleaner air, less food in landfills, and fresher homegrown produce.

Food waste is clearly a major drain on America's local food ecosystems. Addressing it head-on will require government, civil society, and the private sector working in concert to deliver solutions to this multi-faceted issue, mostly on the local and regional levels. Local governments can expand compost programs, support local food banks and food pantries, and proliferate information on how to participate in those programs to everyday citizens. State governments can broaden Good Samaritan laws to include more prepared foods from local businesses that could otherwise go to waste and introduce and clarify regulations on what businesses can trash, donate, or compost when it comes to unsold food. The federal government also can reform expiration dates on food to ensure less is wasted due to unclear labeling. The private sector can play its part by investors funding more innovative startups that transform food waste into valuable products and by food sellers expanding the types of "misshapen" produce they are willing to stock. In this case, a farm on every corner can take the form of a food pantry, a compost bin, or Paula Schafer driving around a truck full of beef. Wasting less can be just as powerful as producing more and can potentially help even more Americans maintain the food security they need to be productive members of their communities.

CHAPTER EIGHT

FARM BILLS AND DOLLAR BILLS: FINANCING SMALL-SCALE AGRICULTURE

Growing up on his small family farm on Maryland's bucolic Eastern Shore, Dan Miller saw first-hand the toll big agriculture had taken on local producers who were trying to farm the land according to their values. It was simply impossible for them to compete with the prices of the factory farms, inputs were becoming more expensive and funding opportunities were limited. In the early 1990s, his parents' farm, like so many others in the region and across the country, was forced to close its doors. Dan moved away from his hometown to Washington, DC to work in real estate, and his family farm remains shuttered to this day. This is a story all-too-familiar to many of America's small farmers who now suffer from the highest debt-to-income ratio the industry has seen in three decades. Competing successfully with the big players while farming according to one's values is just not economically viable anymore for the majority of small-scale agriculturalists.

In a sense, they are being punished for doing right by their local communities and environment. Communities have been forced to surrender their ownership of food production due to the way our government and our society have allowed the agriculture economy to be structured.

Financing A Small Farm

One of the major roadblocks facing small and medium-sized American farms is our economy's agricultural funding ecosystem. The vast majority of funding is driven by government policy, namely the infamous Farm Bill, which is concerned chiefly with the macroeconomics of the agricultural industry often at the expense of the environment and local communities. Beginning in 1938, Congress decided to combine its hodge-podge of farm assistance programs promulgated during the New Deal era into one massive omnibus bill (the most recent of which cost a whopping $857 billion) that was to be updated every five years. It has since grown to encompass a huge number of initiatives like nutrition programs, conservation titles, trade assistance programs, research, credit programs, and even the Food for Peace Act which provides humanitarian assistance in the form of food aid and technical assistance to farmers around the world. Following the passage of NAFTA in 1994, it is designed to support massive corporate farms growing a limited number of commodities for export, such as soybeans and corn. As a result, the Farm Bill provides billions of dollars in subsidies to farms that don't need the money. In fact, top ten farm subsidy recipients each received an average of $18.2 million per year in from 2008 to 2017.[35] In 2017 alone, just four hundred firms picked up between $1 million and $9.9 million each in federal farm subsidies, most of them operating on massive scales as corporate farms.[36] It's also worth noting these farms are growing commodities instead of food, which might be great for their shareholders, but do little to promote national or even regional food sovereignty. This leaves the vast majority of small farmers out to dry in a fiscal environment created not for them, but by the federal government for the big guys.

Unfortunately, in the private financial markets, funding is almost nonexistent for these small producers. In lending markets, most small farmers do not qualify for traditional production credit. This is chiefly due to the inherent risks associated with small-scale farming such as not having sufficient collateral in the form of agricultural land. If they can get a loan, the interest rates may be so high as to be completely prohibitive. This is even more salient for urban farms that may be farming on less than an acre. From a lender and investor perspective, farms require a significant amount of startup capital, are risky low-return investments, and don't offer as predictable a model as other businesses. The risk these small enterprises face is heightened by the trouble of obtaining crop insurance. Only a handful of major crops can be covered under multiple peril crop insurance, that is, a policy that bundles together several different coverages,

35 Adam Andrzejewski, "Mapping the U.S. Farm Subsidy $1M Club," *Forbes*, August 14, 2018.
36 Ibid.

such as for freezes, fires, droughts, and so on. Covered crops are mandated at the county level—if your county doesn't cover what you're growing (as is the case with many small farms), you'd better switch your plantings, or you're out of luck. Unfortunately, affordable credit and accessible crop insurance are simply not options for farms practicing regenerative agriculture or other production methods, and for farms whose markets are regional to hyperlocal.

Also important to understand is most small farmers are not businesspeople. What they know best is farming their land, not the ins and outs of finance and markets. Dr. Theodore Alter is a Professor of Agricultural, Environmental, and Regional Economics and also co-directs the Center for Economic and Community Development at the Pennsylvania State University's College of Agricultural Sciences. He understands this issue on a personal level:

"Most of these small operations are not going to have somebody who's doing the financial side of the business for them. When I look at my own family experience, years ago, we had a greenhouse operation and commercial vegetable operation out in Northwest Ohio. And my brother and I were there for a lot of the labor... my dad was the grower and public relations person, [and] my mother did the accounting and the budgeting and there was no extra help for that. So, I think that managing relationships with the banks and then also navigating the right rules and regulations can be a difficult thing to do."

On top of that, regulations at the local, state, and federal levels, which are written by the powerful corporate farm lobby, can oftentimes be utterly onerous to the small farmer trying to practice a healthy, sustainable agriculture. Many regulations are structured specifically to make business harder for small farmers, requiring them to buy expensive equipment they may not have the funds to acquire. They also may be required to complete thorough, time-consuming regulatory functions, such as those required by the Food Safety Modernization Act (FSMA), which included hundreds of pages of novel safety regulations, many of which have failed to demonstrate their benefits will outweigh their costs.[37] This can include anything from hefty, daily record-keeping requirements to major infrastructure upgrades which can be out of the question for a farm running on thin profit margins. Plus, many states have not yet passed "right to repair" laws that would give farmers and ranchers the ability to fix their own farming equipment. Many agriculture equipment companies have clauses in their contracts that void their product's warranty should the owner attempt to fix it themselves. As a result, farmers might have to take their tractor hours away from their farm to get it repaired at an "authorized repair center," oftentimes for dramatically higher prices than their local mechanic would charge. Farmers end up wasting time and money simply due to burdensome regulations and lack of legal protection.

For urban farms, one must consider zoning regulations for which many small and medium-sized cities have no framework. From a legal standpoint, urban farms are faced with

---

37 Richard Williams, *Regulations Implementing the Food Safety Modernization Act* (Fairfax, Mercatus Center at George Mason University, 2015).

additional challenges with regards to land and water access issues, noise ordinances, and other local laws which may have been promulgated without urban farmers in mind.

The focus here is mostly on legal and food safety regulations and not on Environmental Protection Agency regulations. The latter is more necessary to reigning in the destructive power and negative externalities of big agriculture than the former. Regulations under the guise of food safety can oftentimes be burdensome without much gain for small farmers, such as the FSMA's mandate on recording when your employees wash their hands. Environmental regulations are oftentimes the only thing standing between corporate farms having some semblance of stewardship over their environment and contaminating groundwater with animal waste, rampant pesticide use, and other issues.

A final barrier for small specialty farms is the "knowledge gap." Many folks practicing regenerative farming and raising heirloom varieties of produce and animals enter the profession out of a genuine love and passion for their craft. Best practices for a sustainable business operation may, at best, not always be considered or, at worst, be completely unknown. While corporate farms can afford to hire consultants and MBAs to help them maximize profit by cutting corners where they can't, the small farmer simply does not have the capital to gain access to advanced business practice information. The challenge deepens even further when experimenting with agroecological practices outside of the fertilizer-driven, monocultural mainstream of American agriculture (such as permaculture or Korean Natural Farming) as the common resources available to farmers may not take these methods into account. Thus, they are put at an immediate economic disadvantage against factory farms, even at the local scale.

New Models for Farm Funding

Now back to Dan: he got involved in the new field of crowd-funding, helping folks invest in real estate in their local communities. But when talking to a James Beard, award-winning chef Spike Gjerde and a pioneer of local sourcing who utilizes heirloom varieties from specialty growers, Dan rediscovered the economic issues facing small farmers. Spike lamented the farmers from whom he sourced fresh, unique local ingredients were constantly in danger of insolvency—their lack of access to capital was having a deleterious effect not only on their bottom lines, but also on opportunities for future growth. If their businesses couldn't be sustainable, how could they treat the land sustainably?

Recognizing this pain point led Dan to establish his current venture Steward in 2016, the first "crowdfarming" platform aimed at matching local producers employing sustainable, regenerative practices—oftentimes growing heirloom varieties with local investors wishing to practice their values and give back to the land on which their community rests. Farmers are able to scale-up their operations, upgrade equipment, experiment with new techniques, and more. Local financiers are able to see a return on an investment they know directly helped a member of their local or regional community. This helps communities directly reclaim their food sovereignty by keeping investment dollars in the local economy. This pays dividends not just financially, but also in the increased availability of fresh, local food and a healthier environment. This model offers an incredible way to plug local agriculture into the financial economy of the future and allow them to access capital more affordably and easily than ever before. In addition, Steward helps to address the knowledge gap faced by small farmers by providing them with consulting services and tailored business support to help them succeed.

Models like Steward are the key to bridging both the funding and knowledge gaps with which small farms struggle. It is part of the broader "Slow Money" movement which is about investing consciously in local food systems. In an age of hyper-capitalism, the idea of investors having a responsibility to their communities can seem almost radical. If communities want a shot at reclaiming their food sovereignty, however, they need to prioritize building local networks of private investment to bolster their local producers. Local, state, and federal governments can also create incentives for local investment. These may take the form of tax incentives for investors to dedicate a certain amount of their portfolio to local agriculture. Governments may redirect some subsidies that would typically go to big agriculture and commodities farmers toward creating community development funds and appropriating monies for hyperlocal producers (producers whose consumers tend to live in surrounding communities or even the surrounding blocks). Increasing the dollars earmarked for loans and grants available to small and medium-sized farms as well as farms which practice organic and other regenerative techniques is also essential. Legislative bodies should also take care to repeal some of the onerous regulations that hamper small farmers' abilities to compete with agricultural giants and factory farms in their own communities.

Municipal and city governments have a special responsibility in clarifying their own regulations and zoning laws to make sure urban farmers are not only welcome in their communities, but also supported legally and financially. This is especially important in communities with high food prices and in communities which currently have little to no access to fresh foods. On the federal side, the US Department of Agriculture and Farm Service Agency have a hefty role to play in this effort. They can start by dramatically expanding the dollars available in loans and grants to producers who sell most of their products locally, offering pro bono consulting services as well as business and legal resources to small farmers, and updating their services to be inclusive of small, regenerative farms and specialty crop producers.

All of these proposals represent a brand-new paradigm in how our agricultural system finances and how scarce dollars can be allocated to empower local communities to take ownership of their own food supply instead of toward promoting greater and greater profits by agricultural corporate behemoths. The next decades will see even greater innovation in the financial sector vis-à-vis the concept of Slow Money. This style of responsible investing is centered around "bringing money back down to earth" by incorporating small-scale agriculture into the investment landscape and increasing options for investors to invest locally while being cognizant of their own principles. It is the assertion finance must consider the land's carrying capacity, foster a sense of place, and promote a sustainable care for our commons. This huge and nearly untapped market of hyperlocal investing presents an opportunity ripe for the picking, but it will take a deliberate and value-based approach to ensure this nascent ecosystem is equitable, cooperative, and profitable. Likewise, a novel regulatory scheme must work, first and foremost, for the smallest local players in the market so everyone has a fair shot in the agriculture sector, especially operations with environmental and community commitments.

CHAPTER NINE

TO TREE OR NOT TO TREE: "FOODSCAPING" OUR COMMUNITIES

Growing up in South Central Los Angeles, Ron Finley had long lamented his neighborhood's condition. The racist legacies of redlining in the early to mid-1900s, freeway construction destroying communities, the plant closures and crack epidemic of the 1980s, and recent trends of gentrification and displacement had rendered an area that was once one of the most wealthy and vibrant African-American neighborhoods in the United States into a food desert. Though Ron thought of it less as a food desert and more of a "food prison"—he had to drive forty-five minutes away from his house just to get a fresh tomato! The result is South Central's rates of diabetes, heart disease, and obesity are some of the highest in the region. In fact, less than 10 percent of adults in South Central are consuming the daily recommended servings of fruits and vegetables.[38] All

---

38 Adrian Glick Kudler, "Watts Residents will Die 11.9 Years Before Bel Air Residents," *Curbed Los Angeles*, July 8, 2013.

this culminates in the sad fact residents of the area on average live twelve years less than their wealthier neighbors in West Los Angeles.[39]

So he decided to take matters into his own hands and plant a garden. But Ron didn't just plant a garden: he turned the parkway (the grassy strip between the road and the sidewalk) in front of his house into the area's first food forest. What was once just a plot of dying grass and urban detritus was now bursting with fresh fruits and vegetables. While the City of Los Angeles tried to stop him, claiming he lacked the proper permits, Ron held his ground. Thanks to him, gardening parkways are legal and permit-free for every resident of Los Angeles and Ron's food forests have spread to countless blocks across the South Central. As a self-styled "gangsta gardener," Ron is committed to giving South Central's youth an opportunity to learn how satisfying and cool nourishing yourself with homegrown produce can be. After all, as Ron likes to say: "Growing your own food is like printing your own money!" Now Ron is creating his own nonprofit, The Ron Finley Project, which is currently building its very own urban garden oasis and educational community hub to catalyze the next generation of South Central's gardeners to take up the work.

The story of Ron Finley is just one example of a phenomenon that's occurring around the United States and has gained a lot of traction in the past decade: "foodscaping." Across the country, plenty of towns and cities benefit from tons of empty green space. From parkways, to vacant undeveloped lots, to

---

39 Ibid.

parks in disrepair, there is a tremendous amount of land that can be put to work for the use of all community members. Foodscaping envisions a community ripe with community gardens, fruiting shrubs on the side of roads, food forests in public parks, and other means of building edible horticulture into the fabric of the city.

Foodscaping The City

While to many people think this concept may seem like something of a no-brainer, many cities have long resisted efforts to introduce fruiting plants and trees. Many local governments and planning committees feared such plants would be difficult to take care of and would be a public nuisance if the bulk of the fruit would fall on the ground and end up smashed and rotten. Many municipalities do not want to pay for more city employees to prune trees, apply pesticides, and collect harvests. However, many are now finding there is an army of volunteers who are ready to help out. From the nonprofit Boston Natural Areas Network taking care of orchards all around Beantown (including many in public schoolyards) to Wisconsin's own Madison Fruits and Nuts grassroots working group operating throughout the city's parks, it seems wherever fruit-bearing plants are allowed to be harvested on city property, the community rises up to fill the need. It appears for many cities, the lack-of-volunteers excuse may just be a canard.

The benefits of foodscaping are manifold: from the standpoint of hunger, providing easily accessible fruit-bearing trees in public commons helps local food banks stock up on fresh, local produce. It also helps individuals experiencing homelessness and people in food-insecure situations have a free way to include fresh fruits, vegetables, and nuts in their diets. Moreover, a concerted approach by municipalities to plant fruiting trees and shrubs in neighborhoods which are considered food deserts, that is, an urban area that is located more than a mile away from the nearest supermarket, can do wonders to help alleviate the chronic lack of fresh produce.

Planting trees also has a variety of positive environmental outcomes from absorbing stormwater runoff, to cleaning ground-level ozone and improving air quality, to providing animals with preferable habitat alternatives to obstructing gutters and holing up in attics. A particularly salient example of this is in Hunts Point, a low-income neighborhood in the South Bronx. Hunts Point has the ironic distinction of being a food desert while also being home to the Hunts Point Cooperative Market, the largest food distribution center in the world. The thousands of trucks that rumble their way through the streets every year on the way to the market has caused Hunts Point to suffer from elevated rates of asthma hospitalizations as opposed to other New York City neighborhoods.[40] However, the City of New York in conjunction with local and state organizations lead an effort to plant hundreds of trees in the area, like pear trees, cleaning up and opening new parks and green spaces, and expanding consumer access to fresh produce with mobile fruit and vegetable vendors through their Green Carts project. These and other programs have demonstrably improved health outcomes for Hunts Point residents thanks to weaving trees, nature, and fresh produce into the fabric of everyday life in the area.

---

40 "Hunt's Point Market to Reduce Diesel Fumes in the South Bronx," New York State Office of the Attorney General, June 20, 2003.

Fruiting trees and shrubs can be even more impactful with a dedicated space. Cities all over the nation are embracing a concept called "urban food forests." The idea is to transform what could be a typical city park into a food-producing powerhouse for the local community. Food forests utilize different layers of a forest, such as the canopy, shrub layer, and ground level, to plant fruiting plants that interact with each other in a positive way and contribute to a vibrant ecosystem. Citizens can enjoy a new green space for their outdoor and recreational needs replete with fresh fruits, vegetables, nuts, and sometimes more.

Collaborative Approaches to Foodscaping

One of the most ambitious food forest projects can be found in the Peach State. In line with its citywide goal of 75 percent of residents to live within a half mile of healthy food by 2020, Atlanta, Georgia is piloting its first-ever organic community urban food forest. After going door-to-door for extensive community engagement in the Lakewood neighborhood of Southeast Atlanta, a food desert, they decided upon a site which was once a Black-owned working farm before it shuttered in 2000. Neighbors had fond memories of the Morgan family, who owned the farm, leaving bags of fresh vegetables and eggs on their neighbors' fence posts. The Department of City Planning, in conjunction with a host of government, civil society, and community-based partners, has planted 2,200 new fruit and nut trees on the site with an expected yield of between 440,000 to 880,000 pounds of fresh produce every season after reaching maturity. In addition, the property, now known as the Browns Mill Urban Food Forest, is complete with a medicinal plants garden, a herb garden, a mushroom walk, an apiary for local honey, a working well, and so much more. The site also plans on becoming a cultural hearth for the neighborhood by offering workshops and school trips, potlucks and exercise classes, and a huge variety of other programming. Elizabeth Beaks, a Food Systems Planner with the City of Atlanta, has been thrilled with the project's progress and promise:

"It's the only project I've ever had the chance to work on where our partners are like the food forest itself—We've got federal partners, state partners, local government partners, non-profit partners, local resident partners... and I have never in my life worked with such a passionate, hardworking group of people that are responsive to each other's needs... and with the art, community events, festivals, school trips, [Brown's Mill] is truly one of the most magical places I've ever been in my life."

The ramifications such a space will have on the neighborhood in terms of fresh food access will be stunning, but what's more are the positive externalities of community engagement. The site already has a dedicated cadre of volunteers from around the neighborhood and nearby areas, strengthening community bonds and getting people involved in providing for their own sustenance. Materially empowering a community through food and gardening is an incredibly powerful exercise in uplifting an area, especially economically depressed ones and neighborhoods mostly inhabited by marginalized groups. This is the beginning of a local food economy based on solidarity and engagement instead of on extraction and ought to be the cornerstone of any "green" urban revitalization efforts.

Besides constructing new community gardens and expanding new ones in terms of acreage, yield, and programming, municipalities can do much more to "foodscape" their streets. A concerted effort to plant species that bear fruits, vegetables, and nuts in public parks and greenways is crucial, but it is equally important to get a buy-in from local organizations that will take up the mantle of maintenance with a dedicated corps of volunteers and from local food banks who can use the surplus produce to feed neighbors in food-insecure situations. There are other inventive ways to make use of public produce, such as Atlanta's Ciderfest, a free annual community event with live music where grassroots volunteers serve cider they make from the city's many public apple trees. The possibilities to catalyze community life and serve neighbors in need are truly infinite—all that's needed is a willing municipal government, committed volunteers, and a little hometown pride.

Another way for everyday homeowners to get involved in the foodscaping of their city is starting in their own backyards (and front yards, for that matter.) The modern lawn is a hugely wasteful relic of bourgeoise excess. Suburban homeowners sought to replicate in miniature the huge, stately lawns of European nobility. The United States boasts over thirty million acres of lawns which use roughly sixty million acre-feet of water a year and the lawnmowers who care for them account for a whole 5 percent of the air pollution in the entire United States.[41] What if, instead of wasting so much water, energy, chemicals, and money on manicuring a bed

41 Lakis Carypolou, "The Problem of Lawns," *State of the Planet* (blog) *Earth Institute at Columbia University,* June 4, 2010.

of unsustainable, non-native grass in front of our homes, we turned that energy toward transforming lawns into food-producing powerhouses?

Households could grow more of their own food, nourishing families with quality produce while lowering grocery bills and lowering fertilizer costs and water bills by using native plants. If you think this idea is far-fetched, just take a look back to the Victory Gardens of World War II. Millions of Americans tilled their yards, and at their height, produced between nine and ten million tons of fruits and vegetables, an amount that was equivalent to all the commercially-grown produce available in the nation.[42] This points to an incredible potential just outside our doors that is ripe for the picking. Governments and civil society have a key role to play here: They could promote the foodscaping of private property through tax incentives, send informational mailers, organize free native seed giveaways, facilitate seed exchanges, create citywide composting programs, and roll back restrictions on selling homegrown produce, just to name a few.

Between foodscaped public spaces and private homeowners planting fruits and vegetables in their own lawns and gardens, I have every reason to believe municipalities could meet between 20 percent to 30 percent of their own food needs. This makes foodscaping critical in pursing local food sovereignty, turning our towns and cities into productive members of our food systems rather than just consumers.

---

42 Sarah Sundin, "Victory Gardens in World War II," UC Master Gardener Program of Sonoma County, accessed July 25, 2020.

This food supply would exist mostly outside of the orthodox food economy. The act of procuring your own food for free from your own backyard or from the park or community garden down the street is to declare one's self dramatically more independent from our corporate food system. It's amazing that nourishing one's self and one's community has become something of a revolutionary act, but oftentimes the most radical acts are the most urgent in hindsight.

CHAPTER TEN

RECLAIMING SEEDS AND KNOWLEDGE IN INDIAN COUNTRY

If there's one thing people know about James Kaechele, it's he loves trees. From studying arboriculture in college, to leading the New York Tree Trust, to working with New York City Parks and Recreation to plant one million trees around the five boroughs, he certainly knows his way around an orchard. Yes, when it comes to trees, he has certainly seen it all. That is, until a cherry tree flew right over his head and began its descent into the Grand Canyon.

This was just another day in the programming of the Fruit Tree Planting Foundation (FTPF), a nonprofit that donates fruit and nut trees to communities around the world. In March 2009, the FTPF trained their sights on the town of Supai, Arizona. Supai, the capital of the Havasupai Indian Reservation, is one of the most remote settlements in the nation, located deep in the heart of the Grand Canyon. The Havasupai have a long tradition of orchard-keeping, but the tribe was cut off from over 1.5 million acres of their most fertile land by the federal government in 1882 to carve up for settlers and later for national parks.[43] Due to its location, the town can only be accessed by helicopter, mule, and foot, making the provision of fresh food from the outside world an expensive undertaking. To reclaim their community's food autonomy, the FTPF pledged to make Supai the world's first village in which every family had access to fruit-bearing trees. To make that happen, they flew in over a thousand fruit trees into the canyon by helicopter and planted them in a communal orchard with a team of local and international volunteers. To this day, the inhabitants of Supai enjoy fresh fruits grown in their own backyards and enjoy a level of food sovereignty heretofore unimaginable.

Food Access as a Weapon of Colonialism

While this tale is a heartening example of indigenous people regaining their autonomy in food production, this is unfortunately the exception and not the rule. To address the issue of indigenous food insecurity, we must understand how food access and production has historically been a tool of settler regimes in the Americas. Most people understand the Americas were home to the largest complex of slave-based agriculture in world history. From the Spanish *encomienda* system to the massive plantations of the United States, it was slave labor that fed the booming populations of the New World colonies and made European nations rich with commodities like cotton, tobacco, and sugar. Certain populations

---

43 "Havasupai Tribe," Inter Tribal Council of Arizona, accessed April 26, 2020.

were even selected for enslavement based on their ability to provide food, such as various Mandé tribes who had been growing rice in West Africa for three thousand years and were tasked with growing it in the Low Country of South Carolina.[44] In the United States, capturing runaway slaves and keeping plantations safe from slave revolts is where we begin to see the rise of the modern idea of a standing, bureaucratic police force. Nearly one hundred years before major cities such as New York and Chicago created centralized municipal police departments, Southern colonies developed slave patrols under the auspices of county courts.[45]

Of course, slavery did not truly end with the Emancipation Proclamation. Even after the Civil War, many "freed" slaves continued to be sharecroppers in their former masters' fields. The relationship between the police and slavery continues to this day with prisoners working on government-owned farms across the country, primarily in Arkansas and Texas. It's possible the strawberry you eat today was picked by a prisoner earning just a few cents an hour. In the present-day, American food production is still stratified along racial lines. Many of those at the bottom of the industry working low-wage jobs and contributing hard manual labor are Mexican migrant workers. Furthermore, most of our country's food deserts are located in majority-minority areas with the

---

44 Joseph A. Opala, ""The Gullah: Rice, Slavery, and the Sierra Leone-American Connection," Gilder Lehrman Center for the Study of Slavery, Resistance, and Abolition at Yale University, accessed April 26, 2020.

45 Victor E. Kappeler, Ph.D., ""A Brief History of Slavery and the Origins of American Policing," Police Studies Online at Eastern Kentucky University, accessed April 26, 2020.

hardest-hit areas being urban cores and Indian Country. This lack of fresh food access has a direct impact on the health of Black and indigenous families, in some cases causing dramatic rates of diabetes, heart disease, and obesity.[46]

Due to this history of dispossession, genocide, and subjugation under the heel of European colonialism and the American settler state, the indigenous people face challenges related to food security unmatched by any other demographic cohort. One in four Indian households regularly face food insecurity—more than twice the numbers for white households. Indian children have nearly twice the levels of obesity than other American children. While many residents of Indian reservations (nearly 85 percent in some areas) receive nutrition assistance through government programs like the Food Distribution Program on Indian Reservations (FDPIR). Although nowadays these deliveries include fresh produce, they historically contained mostly packaged and canned foods high in salt, fat, and sugar.[47] It's no surprise American Indians face levels of Type II diabetes nearly three times higher than their non-Hispanic white counterparts.[48] This situation is compounded by the fact many indigenous communities are food deserts without a grocery store.

---

46 Sandra A. Black, "Diabetes, Diversity, and Disparity: What Do We Do with the Evidence?" *American Journal of Public Health* 92 no. 4 (April 2002): 543-548.

47 Mary Kay Fox, William Hamilton, and Biing-Hwan Lin, *Effects of Food Assistance and Nutrition Programs on Nutrition and Health: Volume 3, Literature Review,* U.S. Department of Agriculture, Economic Research Service (Washington, 2004).

48 Elias K. Spanakis, and Sherita Hill Golden, "Race/Ethnic Difference in Diabetes and Diabetic Complications," *Current Diabetes Reports* 13 no. 6 (2013): 814-823.

Individuals living in those towns and reservations must make use of whatever food is for sale in gas stations and convenience stores. Any suite of policy recommendations to build more robust local food systems in our nation need to start with those who are the most direly affected, and perhaps no group faces more deprivations than indigenous Americans.

Reconnecting with Indigenous Foodways in the United States

Indigenous food insecurity is a more nuanced situation than just a history of discrimination, poverty, and food deserts. As the original inhabitants of the land now called the United States, they had a wealth of natural resources from which to choose. Some groups, like the Chumash in what is now Los Angeles County, were hunter-gatherers, reaping the bounties of their local coasts, forests, mountains, and rivers. Others, like the various cultures of the Mississippi valley, were advanced agriculturalists, farming maize, squash, and even sunflowers, forming complex societies with cities like Cahokia rivaling contemporaneous London at their peak.[49] It's impossible to know how many people were living in pre-Columbian America, but scholars agree there were millions of people living North of the Rio Grande until European contact. While hunger was not unknown on the continent, this was typically experienced in the context of famine caused by natural droughts and not by sociopolitical structures.

---

49 ""Greater London, Inner London & Outer London Population & Density History," Wendell Cox Consultancy, accessed April 26, 2020.

Regardless of social structure, these civilizations all cared for the land on which they depended. Across the continent, American Indians were using fire to clear undergrowth and to promote the natural proliferation and intentional growth of plants which bore vegetables, fruits, nuts, seeds, and other edibles. In areas like the Northeast, the Rockies, and California, indigenous agroforestry worked with the land and its ecosystems, allowing nature to invade fallow fields and repopulate burnt areas with nitrogen-fixing grasses and berry bushes which thrived in the nutrient-rich, ashy soil. This symbiotic relationship between nature and food production turns Western conceptions of agriculture on its head—there was no plowed field encroaching on an untamed wilderness. Rather, nature itself became their harvest and every aspect of their "wilderness" was very carefully maintained. This native stewardship of the land worked well for centuries, allowing populations to grow with the land's natural carrying capacity and guiding nature toward a more bountiful ecosystem.

Of course, this relationship between humans and their environment in North America changed with the arrival of Europeans on the continent. They carried diseases to which the Indians had no immunity, like smallpox and measles, decimating their populations by upward of 80 to 90 percent.[50] Some areas saw the wholesale death of tribes, rendering them completely uninhabited. Entire tribes were enslaved to labor in mines, in homes, and on plantations—plowing the land to which they had once belonged. Native ways of life were forcibly and inexorably altered as they were boxed into

---

50 Noble David Cook, *Born to Die; Disease and New World Conquest 1492-1650* (Cambridge: Cambridge University Press, 1998), 1-14.

smaller and smaller reservations, coerced into signing unfair and often cruel treaties. The effect this had was disastrous: Native land management practices were replaced with Western agricultural methods which did not work with the land. Instead, its purpose was to extract as much crop as possible for the commodities trade and to feed both the burgeoning colonies as well as their European overlords. This sudden transition and destruction has had extremely harmful effects on the environment of the Americas, the effects of which are still being seen today. The unsustainable farming the Europeans introduced has caused the erosion of topsoil across the country, leading to some places physically sinking by multiple yards in height over the past few hundred years. Even today, the United States loses nearly three tons of topsoil per acre every year and more and more acres of farmland become infertile. These unsustainable practices have destroyed much of the land on which American Indians were able to grow and harvest their own food.

Moreover, many indigenous groups lost most if not all of their land through the treaty system, a classic example being the Trail of Tears, where forty-six thousand people were ethnically cleansed from their homelands in the Southeastern US and relocated to what is now part of Oklahoma.[51] This dispossession meant indigenous people had to try and relearn agriculture in an unfamiliar land with unfamiliar crops. This, coupled with the destruction of many indigenous oral histories under colonialism, forced conversions to Christianity, kidnappings, and more, meant new generations were

---

51 ""Indian Removal," Africans in America, Public Broadcasting Service, accessed April 26, 2020.

unaware of much of their people's history. This knowledge included not just belief systems and folklore, but practical agricultural and culinary skills. Much of the knowledge of how their ancestors grew and prepared food has been lost. What has been recovered is difficult to disseminate to those who have been left with few economic resources or fertile land. How can a person be expected to feed themselves when they have been deprived of the recipe, the kitchen, and the garden?

This situation is compounded by the loss of biocultural heritage in the United States. Many indigenous varieties of plants have been replaced by just a handful of mass-produced species. Heritage breeds of crops stand little chance against the onslaught of supermarket-approved choices that are compatible with our modern agricultural and retail system. Many seeds that were commonly available to American Indian tribes just a few hundred years ago are now extinct or unknown to their descendants. Furthermore, major seed companies, like the aforementioned Monsanto, have a vested interest in replacing indigenous varieties with their patented seeds, sacrificing biodiversity for short-term profits. All in all, the relationship between food and American Indian communities is multi-faceted. One marred by land dispossession, knowledge gaps, lack of economic power, and a loss of biological heritage.

Regaining American Indian Food Autonomy

The good news is many proud Americans of indigenous descent are working tirelessly to address their own community's needs. One such individual is Chef Sean Sherman of the Oglala Lakota. Growing up on his tribe's impoverished

Pine Ridge Indian Reservation in South Dakota, Sean started working in kitchens in his early teenage years to provide for his family. For years, he studied diverse cuisines from around the world: from Japanese, to Mexican, to the perennially popular French cuisine. However, Sean realized the vast culinary traditions of indigenous Americans were never discussed. He realized how relatively little of his own Lakota culture's food and cooking he knew. This deprivation hinted at more than just an accidental slight against his heritage by the Western culinary world, but rather a complete abdication of responsibility by American chefs to promote the autochthonous food of the land on which they cooked. But the problem ran deeper still: there were very limited resources on American Indian foodways, and those that existed were inaccessible to the average chef or farmer.

Thus, Sean took it upon himself to breathe light into this forgotten culinary world, working with ethnobotanists, food historians, tribal leaders, and countless others. Sean rediscovered the traditional crops and foods of what is now the Dakotas and Minnesota and composed his own cookbook: The James Beard award-winning "The Sioux Chef's Indigenous Kitchen." This landmark publication was a completely novel way at exploring American cuisine. He did not employ any colonial ingredients like wheat, dairy, sugar, or beef in favor of native wildflowers, sumac, blueberries, lake trout, and more. His recipes introduced healthy American Indian foods into the modern culinary dialogue for the first time, but he's not done yet. With a catering company, a food truck in the Twin Cities area, and plans to open a modern Lakota restaurant in Minneapolis, Sean is bringing his culture's edible treasures to the American plate. But Sean also understands his mission is about more than just profit or reconnecting with his own personal heritage. He intends to open "indigenous food labs" around the country and world to help native youth and adults discover their own unique culinary legacies and economically empower native communities. He asserts that not just those in Indian Country can benefit:

"You don't have to be indigenous to appreciate the benefits of indigenous diets and to understand the true indigenous histories of the region and the land that you're currently residing on. We can utilize and our indigenous communities to be role models for the rest of the world... because if we can help people and tribes who have suffered so much to regain a lot of their food security themselves and to really start to work on [indigenous food production] systems that are really effective, then we can showcase how we can do this everywhere."

Chef Sean Sherman foraging Labrador for use in a Lakota recipe

Grassroots approaches like Chef Sean's are the key to reviving food sovereignty in Indian Country. Not only is his approach a bottom-up paradigm that involves disseminating knowledge and hands-on experience directly into communities, but it also involves building livelihoods based around cultural heritage to preserve and propagate it not just on reservations, but into mainstream American society. By putting this untapped market into the hands of those to whom it rightly belongs, it will essentially create its own demand as more and more people explore indigenous culture and uplift Indian communities through enterprise. More money means more opportunities for indigenous communities to build their food autonomy on their own terms.

Running parallel to job creation is the power of indigenous land management and saving heritage breed crops. A multitude of wonderful organizations around the world are fighting to keep land and seeds in native hands. The Native American Food Sovereignty Alliance operates an Indigenous Seed Keepers Network to educate native farmers on the use of heritage seeds, repatriate seeds from seed banks to their respective tribes of origins, and organize advocacy campaigns and events to coalesce the community behind actionable goals and to support local and tribal seed-saving initiatives. The Traditional Native American Farmers Association focuses on educational programs to revitalize traditional indigenous methods of land management and agriculture based on the axiom "healthy soils equal healthy plants equal healthy people equal healthy Nations." The Indigenous Food and Agriculture Initiative (IFAI) out of the University of Arkansas tackles the unique legal challenges faced by tribes as they build healthy food systems through strategic legal analysis, policy research, and educational resources. Using training webinars for Indian-owned food businesses, comprehensive economic forecasting and food systems reporting, and lobbying for a "Native Farm Bill" to address these disparities at the federal level, IFAI brings academic expertise in reach of the entire native food systems community. Such a multidimensional approach from all these stakeholders is key to addressing the multiple deprivations faced by the American Indian community.

Bringing a farm to every corner (and canyon) of Indian Country will not be easy, but it is one of the most consequential and powerful initiatives one can support in uplifting our indigenous brothers and sisters. Through empowering native communities with the economic tools, generational knowledge, and government policies that will support their families and businesses, every American can help build healthier, wealthier, more culturally enriched indigenous communities.

PART THREE

REIMAGINING THE AMERICAN FOOD SYSTEM

CHAPTER ELEVEN

A SEAT AT THE (FARM-TO-) TABLE: BUILDING AN INCLUSIVE FUTURE AND CLAIMING OUR POWER

Now that we are one-fifth of the way through this century, it is becoming more and more evident the problems of the next eighty years and beyond will require a fundamental reorganization of our society. Across all sectors, "business as usual" simply cannot continue on at the same pace or within the same antiquated structures. Minor tweaks or symbolic changes on the behalf of governments and corporations are soon becoming an embarrassing patchwork of small token projects, grand but ultimately insignificant philanthropic overtures, and virtue signaling. These gestures are emblematic of the failure of the neoliberal world order to properly respond to the various crises which plague the system. The solution to the looming threat of climate change on American agriculture? Better double-down on corporate farms and CAFOs to make as much in short-term profits as they can before time runs out. What is the solution to food security during the COVID-19 pandemic? Let our massive agricultural supply lines get disrupted and watch on the evening news as farmers destroy thousands of pounds of uneaten produce in their fields. These blows to our system will only worsen over the next few decades and compound with other issues. With regard to our food, we need to understand the limits on our current system's ability to adapt and propensity for future growth will seriously damage our society's resilience in the face of systemic shocks. We must examine all of the obstacles the American food system will face in the coming years to weather them.

The Coming Challenges of the Twenty-First Century

By 2100, we will have officially run out of the vast majority of our planet's arable topsoil, the top layer of soil that is essential for all plant life, and by extension, the entire natural world. Our unsustainable agricultural practices degrade soil anywhere from ten to forty times the rate at which it can be naturally replenished.[52] This global catastrophe will gravely impact the ability of humankind to produce the food we will need to nourish our growing population. Moreover, it will dramatically reduce agricultural productivity in the long run.

---

52 David Pimental, "Soil Erosion: A Food and Environmental Threat," *Environment, Development, and Sustainability* 8, no. 1 (February 2006): 119-137.

The food security of communities will rely heavily on the protection and remediation of their depleted soils. Moreover, the adoption of alternative, no-soil modes of agriculture, such as indoor hydroponic and aeroponic systems, cannot remain in the realm of novelty. These solutions must be embraced as critical elements of local and regional food systems so as to decouple local supply chains from deteriorating soils and unproductive agriculture across the country and the world.

As the century progresses and global warming continues to wreak havoc in the United States and around the world, billions of people will not have access to a clean, reliable supply of potable water. Lack of fresh water does not only pose a serious concern to human populations. This is also intrinsically linked to the vitality of our agriculture. Many communities that already suffer from regular drought, such as the Central Valley in California, where 80 percent of water pumped by humans was used in farming and ranching.[53] As the world gets drier and hotter, rainfall will become progressively sporadic. This trend becomes even more problematic when communities around the country, especially those which have ample access to fresh water, must rely on the rainfall in a small handful of breadbasket locales for their food security. Reclaiming control of their agriculture becomes a matter of survival. Bringing food production closer to home can help insulate communities from shocks suffered by the distant farms on which they once relied. Furthermore, utilizing agricultural practices which conserve water, such as drip irrigation, rotational grazing of livestock, farming

---

53 Jeffrey Mount and Ellen Hanak, *Just the Facts: Water Use in California* (San Francisco, Public Policy Institute of California, 2019).

with the natural (or purposely engineered) geography of the land to maximize rainfall usage by crops and to minimize runoff, and proper soil management and deep tilling can create a network of regenerative farms which work together to preserve a region's precious watersheds from pollution and depletion. These farms will continue to provide fresh, nutritious (and indeed, water-filled as they would have less time between farm and plate to lose moisture) food to nearby markets no matter what happens in traditional breadbasket regions.

A New Deal for Local Food

The promise of eternal economic growth accompanied by an utter disregard for our environment and frontline communities will come to a veritable impasse in the coming century with increasingly scarce resources. Now is the time for governments over the world to put forward meaningful, comprehensive legislation to address these challenges. Here in the United States, proponents of the "Green New Deal" recognize the inflection point at which we are quickly arriving and have put forth an assortment of urgent, transformative solutions to limit American contributions to global greenhouse emissions and provide relief for those whose jobs will be eliminated by those proposed changes. Section 2G of the House Resolution 109 "Recognizing the duty of the Federal Government to create a Green New Deal" speaks on our country's pressing need to pursue a better path forward for our food system and calls for:

"supporting family farming... investing in sustainable farming and land use practices that increase soil health; and... building

a more sustainable food system that ensures universal access to healthy food."[54]

Congresswoman Alexandria Ocasio-Cortez unveiling the Green New Deal on Capitol Hill

The Green New Deal is a laudable effort from a few members of a government in which members from both major parties are typically paid off by corporate agriculture lobbyists. Unfortunately, a vague, nonbinding resolution simply does not go far enough to address the massive systemic issues this country is facing in terms of its food supply. What is needed is a New Deal-type program to rapidly and radically transform our agricultural, food distribution, and food waste paradigms before it is too late. A

---

54 U.S. Congress, House, *Recognizing the duty of the Federal Government to create a Green New Deal* HR 109, 116th Cong., 1st sess., introduced in House February 7, 2019.

"New Deal for Local Food" would harness the human and capital resources of the world's most powerful government to create a national system of regionalized food systems which would serve to bolster American food security, fight climate change, protect our ecosystems, feed our growing population, improve the health and well-being of our citizens, protect marginalized communities with special attention to our indigenous communities, support small businesses and local economies, and promote a more flavorful American gastronomy.

An all-encompassing approach would realistically take years and cost billions of dollars to achieve. It would require a monumental effort on the part of federal, local, state, and tribal governments, businesses of all sizes, and civil society. The reordering of our agricultural practices along small-scale, local, and regenerative lines can take place through a system of taxes and incentives, shifting the landscape to where producers who take care of their land, grow native crops or heritage breeds, and sell their products locally are supported while corporate agriculture is dramatically disincentivized from exercising their destructive, unsustainable practices. Laws would have to change to make it easier for small producers to sell their goods in their communities, for foodscaping to become an integral part of municipal planning, for cities to permit and support urban agriculture, and for average citizens to raise small livestock in their home and backyards. Local and state budgets would also have to be reprioritized to support community gardens; food hubs; farmers markets; NGOs that deal with food access; and food pantries, food banks, and the resource-sharing networks thereof.

The USDA would have to take the lead on providing no-interest lines of credit and insurance to producers who are selling to local markets, growing produce and heritage seeds, and are employing responsible land management practices. Their research dollars should be spent on developing free-to-use crop varieties that need less water and grow faster and breeding animals that produce less greenhouse emissions while yielding more meat. They can also reposition themselves to be a leader in providing farmers with the resources they need to retool in learning regenerative agricultural practices as well as educating the public on the importance of regionalizing our food systems and supporting your local farmers and ranchers.

Restaurants must make greater use of local foods in their kitchens and adopt a seasonal approach to menu development. They can also be local leaders in food waste management and support local composting initiatives. Funding to the Bureau of Indian Affairs should be dramatically increased so they can respond effectively to the crises surrounding food security in Indian Country. The Bureau's mandate should be expanded to include the building of tribal food sovereignty, the promotion of traditional agricultural and culinary practices in the modern economy, the dissemination and saving of heritage seeds, and access to no-interest loans for native-owned businesses.

Building Dual Power

It would also require a lot from all of us, too, as citizens of our respective regions. We would need to ask ourselves: what am I willing to sacrifice in the short-term to secure a better tomorrow for future generations? Am I prepared to swap "fresh" fruits in wintertime for frozen or canned ones? Am I ready for meat to cost more and to eat less of it? Do I have the dedication to compost my food scraps and engage with local collection programs? Can I put in the time and effort to plant a fruiting tree or raise some chickens in my backyard? Am I willing to volunteer with my community garden or farmers market? Am I willing to change my spending habits to prioritize local and minority-owned producers? No one has to do all of these things to the umpteenth power, but the truth is everyone will have to take part in some of them—and then some. Only then can the structural transformation of our food economy, agriculture, and governmental programming and regulatory schema have the profound impact on our lives and behaviors it needs to have to be effective.

An even bigger question is how to get rural America on the side of change. Those who work and live off the land oftentimes have the greatest sense of stewardship for the earth. They are also the folks who will be most affected by climate change and by the dramatic changes which a New Deal for Local Food would entail. However, decades of lobbying and politicking have convinced some that these sorts of urgent solutions are the exclusive property of the political Left. At the same time, many urban liberals and progressives are convinced rural Americans are unreachable or unwinnable for their cause at best, or hostile to progress at worst. The truth is these issues are not a left versus right issue, but a top versus bottom issue. Rural people know they have been relegated to the bottom of the political priorities list for far too long and their real concerns around their land and livelihoods are too often abandoned by both major parties. The hardworking folks who make up our rural communities are hyperaware of the effect big corporations have had on their way of life and they understand all too well the dangers of a changing climate on their farmland. A broad coalition of country folks and city-dwellers, those who the Reverend Jesse Jackson coined the "eaters and the feeders," is not only possible, but critical to building a massive grassroots movement to save America's small farms and transition to regenerative agricultural practices. This can only be done by listening intently and speaking directly to the values of those who call our countrysides home to work together without regard to race, party affiliation, location, and any other label. We are all in this together and we need to start acting like it.

Our society stands at a pivotal point in time—the behemoth of big agribusiness and their benefactors in our government seems unstoppable. As our environment degrades, communities fall apart, and small farmers go bankrupt, big agribusiness enjoys record profits and an unprecedented command of our political arena. To achieve the world we know is possible, it will take tireless organizing at the community level to achieve results in government. But it will also take ordinary people who are committed to the idea of building "dual power" with their neighbors. This means communities who value their food sovereignty will have to step up to the plate and create institutions which will do the good work governments and big agriculture refuse to do. This can already be seen in food hubs, nonprofits promoting food security in underserved areas, and other organizations that are moving the needle away from corporate control of our food toward more democratic, equitable food systems. Ultimately, dual power seeks to usurp the state and allow communities to have complete control over how their food is produced, how the land it was produced on is managed, how it is distributed, how the labor went into producing the food is treated, and how food waste and the byproducts of production are handled. In this way, people need not wait for crumbs at the table of the powerful—they can claim the power for themselves. If we are to be truly serious about a farm on every corner—or a food hub, a community garden, a farm-to-fork restaurant, a farmers market, and so on—it is incumbent on all of us to be the change we so desperately want to see.

ACKNOWLEDGMENTS

First and foremost, I would like to thank my wonderful family. To my mother and father, I know how proud you would be of your son publishing his first book. Thank you for raising me to work hard and follow wherever my passions lead me. To Schlübbo, Emily, and Molly, you are the best and most supportive siblings I ever could have asked for. To Evelyne, thank you for always believing in me and loving me like your own child. To Annie and Adam, thank you for always loving and being there for me and my siblings during some of the most difficult times in our lives. To Livia, thank you for being the most wonderful (and only) cousin a guy could ever want. To Linc and Soiz, thank you for being such kind and warm "in-laws" to me and bringing us into the Caplan fold. Thank you to Dadi, Michal, and Ilan for keeping the family line strong in the Holy Land. Todah rabah!

To the Cheney family, I am so grateful to have you all in my life. Grace, Brian, Connor, Sean, Nick, Brendan, Ollie, and Chester—you have all become like a second family to me over the years. Thank you for letting me steal Sean away whenever I want and for having me as the honorary fifth Cheney boy.

To my Georgetown colleagues and friends Ashleigh, Ben, Brooke Callie, Eric, Jeff, JP, Megan, and Liv, you all are the reason I'm proud to be a Hoya. Working and studying at the Hilltop is only worth it because you all have left your marks there and I'm all the better for it.

Thank you to all the Delta Lambda Phi fraternity brothers who have supported me on my authorship journey. To Aidan, Amit, Bryan, Cam, Connor, Henry, James, Jonathan, Josh, Kevin, Kyler, Luke B., Luke P., Wade, and Xavier, strong the circle we! You all make me proud every single day to be a Lambda Man.

To my George Washington University friends Lauren, Claire, Aaditya, Jenevieve, and Jack, always Raise High! Thank you for enriching my undergraduate experience and setting me up for success in the real world.

To Samantha and the McCallion family, thank you for the unconditional support you give me no matter what insanity I'm partaking or wherever you are in the world. I know I can always count on you all.

To the Van Ettens for always treating the Langes as a part of the family. I know our families will always be there for each other no matter what.

To all my Scotch Plains and Fanwood friends—you all made me who I am today. Thank you to Brooke, Becca, Amanda, Emma, Liz, Cori, Garrett, Brianna, Alyssa, the Sheehy family, the Zikas family, the Tice family, the Pannuri family, the Walejewsky family, and the Iannacone family. Thank you to my grade school teachers Mr. Stack, Mrs. Brodsky, and Mrs. Allen for inculcating in me the skills, knowledge, and work ethic I needed to take on writing my first book.

Special thanks are owed to Julian and Garret, with whom I have lived with for more than two years. Thank you for being my friends, my confidants, my roommates, my rocks, and for Julian, my Big Brother.

Thank you to all my interviewees, without you this book would not be possible. I am very grateful to Aaron de Long of Pasa Farming, Alexis Russell of Ayrshire Farm, Autumn Rae Ness of Maui Food Hubs, Elizabeth Beak of the City of Atlanta Office of Resilience, Dan Miller of Steward, Eric Pederson of Ideal Fish, Jeremy Kranowitz of Keep Indianapolis Beautiful, Inc., Jonathan Bethony of Seylou Bakery, Liz Warren-Novick of 80 Acres Farms, Marbury Jacobs of The Garden Farmacy, LLC, Marc Oshima of AeroFarms, Nicolas Jammet of Sweetgreen, Nicholaas Mink of Sitka Salmon Shares, Paula Schafer of MontCo Anti-Hunger Network, Samuel Thayer of Forager's Harvest, Sean Sherman of The Sioux Chef, Dr. Ted Alter of the Pennsylvania State University College of Agricultural Sciences, and Will Horowitz of Ducks Eatery. I am deeply grateful for you all sharing your wisdom, stories, and love of food with me. This book is a testament to the rich diversity and warm heart of our industry.

Thank you to my publishing team at New Degree Press. To Jemiscoe and Morgan, it was an honor to have my writing undergo your critical review and I am so pleased to have had you both as a part of my authorship journey.

Thank you to all others who were so kind as to preorder this book, such as Alex, Carolyn, Cory, Dan, Danielle, DiAnna, Elissa, J-Cass, Jennifer, Kiersten, Kristen, Peter, and Sa'ad. You all helped make this possible and I am extremely grateful for your faith in me.

APPENDIX

A Song of Sprouts and Steel

- Pew Research Center. "U.S. Population Projections: 2005-2050." Last modified 11 February 2008. https://www.pewresearch.org/hispanic/2008/02/11/us-population-projections-2005-2050/.

- Stocker, T.F., et al. *Climate Change 2013: The Physical Science Basis. Contribution of Working Group I to the Fifth Assessment Report of the Intergovernmental Panel on Climate Change.* Cambridge: IPCC, 2013. Accessed April 20, 2020. https://www.ipcc.ch/site/assets/uploads/2017/09/WG1AR5_Frontmatter_FINAL.pdf.

A Brief History of American Agriculture and Food Sovereignty

- Berry, Wendell. *The Unsettling of America: Culture & Agriculture.* San Francisco: Sierra Club Books, 1977.

- Fox, Matthew. *A Way To God: Thomas Merton's Creation Spirituality Journey.* Novato: New World Library, 2016.

- Hornbeck, Richard. "The Enduring Impact of the American Dust Bowl: Short- and Long-Run Adjustments to Environmental Catastrophe." *American Economic Review* 102, no. 4 (June 2012). https://www.aeaweb.org/articles?id=10.1257/aer.102.4.1477.

- Oklahoma State Legislature, Senate, "A Concurrent Resolution Designating April 14, 2015 as 'Dust Bowl Remembrance Day,'" S 16, 55th Oklahoma Senate., 1st sess., adopted in Senate April 13, 2015.

- Pack, Charles Lathrop. *War Gardens Victorious*. Philadelphia: J. B. Lippincott, 1919.

- "The Great Depression Hits Farms and Cities in the 1930s," Iowa Public Broadcast Corporation, accessed Aug 22, 2020. http://www.iowapbs.org/iowapathways/mypath/great-depression-hits-farms-and-cities-1930s.

Little Farm, Big Impact: Natural-Scale Local Agriculture

- Stannard, David E. *Before the Horror: The Population of Hawai'i on the Eve of Western Contact,* Honolulu: Social Science Research Institute, University of Hawaii, 1989.

- U.S. Department of Agriculture, National Agriculture Statistics Service, *2007 Census of Agriculture*. Washington, D.C., 2009. https://www.nass.usda.gov/Publications/AgCensus/2007.

A Farm on Every Corner: Agriculture in the Big City

- Abundance North Carolina. "Breaking Barriers: How Urban Gardens Impact Crime." Accessed June 20, 2020. https://abundancenc.org/breaking-barriers-how-urban-gardens-impact-crime/.

- Bellows, Anne C. Food Security Dot Org. "Heath Benefits of Urban Agriculture Public Health and Food Security." Accessed June 20, 2020. http://foodsecurity.org/uahealthfactsheet/.

- Poulsen, Melissa, Roni Neff, and Peter Winch. "The multifunctionality of urban farming: perceived benefits for neighbourhood improvement." *Local Environment* 22, no. 11 (2017). https://doi.org/10.1080/13549839.2017.1357686.

- Terry, Natalie and Kara Gross Margolis. "Serotonergic Mechanisms Regulating the GI Tract: Experimental Evidence and Therapeutic Relevance." *Gastrointestinal Pharmacology. Handbook of Experimental Pharmacology* 239, (2016). https://dx.doi.org/10.1007%2F164_2016_103.

- *The Place of Urban and Peri-Urban Agriculture (UPA) in National Food Security Programs.* Rome: Food and Agriculture Organization of the United Nations, 2011. Date accessed June 20, 2020. http://www.fao.org/3/i2177e/i2177e00.pdf.

The Art of the Veal: Bringing Livestock Home

- *Fish to 2030: Prospects for Fisheries and Aquaculture.* Washington: The World Bank, 2013. http://www.fao.org/3/i3640e/i3640e.pdf.

- Mullhollem, Jeff. "Feed Supplement for Dairy Cows Cuts Their Methane Emissions by About a Quarter." Penn State Department of Agricultural Science· Accessed April 18, 2020. https://animalscience.psu.edu/news/2020/feed-supplement-for-dairy-cows-cuts-their-methane-emission-by-about-a-quarter.

- Simon, David Robinson, *Meatonomics: How the Rigged Economics of Meat and Dairy Make You Consume Too Much.* Newburyport: Conari Press, 2013.

- Stahler, Charles. The Vegetarian Resource Center. "How Many People Are Vegan? How Many People Eat Vegan When Eating Out? Asks the Vegetarian Resource Center," Date accessed April 18, 2020. https://www.vrg.org/nutshell/Polls/2019_adults_veg.htm.

- "The United States Meat Industry at a Glance," American Meat Institute, accessed April 18, 2020. https://www.meatinstitute.org/index.php?ht=d/sp/i/47465/pid/47465.

- U.S. Department of Agriculture, National Agriculture Statistics Service, *September 2020 World Agricultural Supply and Demand Estimates*. Washington, 2020. https://www.usda.gov/oce/commodity/wasde/wasde0920.pdf.

- U.S. Department of Agriculture, National Agriculture Statistics Service, *Honey*. Washington, 2018. https://www.nass.usda.gov/Publications/Todays_Reports/reports/hony0318.pdf.

Wild Salmons and the Limits of Localism

- *Alaska Seafood Export Market Analysis*. Juneau: Alaska Seafood Marketing Institute, 2016. https://www.alaskaseafood.org/wp-content/uploads/2015/10/Spring2016-Alaska-Seafood-Exports-Final.pdf.

Restaurants at the Human-Scale

- Davies, Terry and David M. Konisky. *Environmental Implications of the Foodservice and Food Retail Industries*. Washington: Resources for the Future, 2000. https://doi.org/ 10.22004/ag.econ.10761.

- Hales, Craig M., MD, et al. Center for Disease Control, National Center for Health Statistics, *Prevalence of Obesity and Severe Obesi-*

ty Among Adults: United States, 2017–2018. Hyattsville, Maryland, 2020. https://www.cdc.gov/nchs/data/databriefs/db360-h.pdf.

- Riehle, Hudson, et al. *Restaurant Industry 2030: Actionable Insights for the Future.* Washington: National Restaurant Association, 2019. Restaurant.org/Restaurants2030.

How the (Veggie) Sausage Gets Made: Addressing Food Waste from All Sides

- Aggidis FIMechE, George, et al. *Global Food: Waste Not, Want Not.* London: Institution of Mechanical Engineers, 2013. https://www.imeche.org/policy-and-press/reports/detail/global-food-waste-not-want-not.

- Baker-Smith, Christine, et al. *#RealCollege 2020: Five Years of Evidence on Campus Basic Needs Insecurity.* Philadelphia, The Hope Center for College, Community, and Justice at Temple University, 2020. https://hope4college.com/wp-content/uploads/2020/02/2019_RealCollege_Survey_Report.pdf.

- Bartoleschi, Pietro. *Food Wastage Footprints.* Rome: Food and Agriculture Organization of the United Nations, 2013. http://www.fao.org/3/i3347e/i3347e.pdf.

- Gunders, Dana. *Wasted: How America Is Losing Up to 40 Percent of Its Food from Farm to Fork to Landfill.* New York City: National Resource Defense Council, 2012. https://www.nrdc.org/sites/default/files/wasted-food-IP.pdf

- Gunders, Dana and Jonathan Bloom. *Wasted: How America Is Losing Up to 40 Percent of Its Food from Farm to Fork to Landfill.* New York City: National Resource Defense Council, 2017. https://www.nrdc.org/sites/default/files/wasted-2017-report.pdf.

- *The Impact of the Coronavirus on Food Insecurity*. Chicago: Feeding America, 2020. https://www.feedingamerica.org/research/coronavirus-hunger-research.

- Wieben, Emilie. *Save Food For a Better Climate: Converting the Food Loss and Waste Challenge into Climate Action*. Rome: Food and Agriculture Organization of the United Nations, 2017. http://www.fao.org/3/a-i8000e.pdf.

Farm Bills and Dollars Bills: Financing Small-Scale Agriculture

- Andrzejewski, Adam. "Mapping the U.S. Farm Subsidy $1M Club." *Forbes*, August 14, 2018. https://www.forbes.com/sites/adamandrzejewski/2018/08/14/mapping-the-u-s-farm-subsidy-1-million-club/#2effcf473efc.

- Williams, Richard. *Regulations Implementing the Food Safety Modernization Act*. Fairfax: Mercatus Center at George Mason University, 2015. https://www.mercatus.org/system/files/Williams-FSMA-Regulations.pdf

To Tree or Not to Tree: "Foodscaping" Our Communities

- Carypolou, Lakis. "The Problem of Lawns." *State of the Planet* (blog) *Earth Institute at Columbia University*. June 4, 2010. https://blogs.ei.columbia.edu/2010/06/04/the-problem-of-lawns/.

- Kudler, Adrian Glick. "Watts Residents will Die 11.9 Years Before Bel Air Residents." *Curbed Los Angeles*. July 8, 2013. https://la.curbed.com/2013/7/8/10222942/watts-residents-will-die-119-years-before-bel-air-residents.

- New York State Office of the Attorney General. "Hunt's Point Market to Reduce Diesel Fumes in the South Bronx." June 20, 2003. https://ag.ny.gov/press-release/2003/hunts-point-market-reduce-diesel-fumes-south-bronx.

- UC Master Gardener Program of Sonoma County. "Victory Gardens in World War II." Sarah Sundin. Accessed July 25, 2020. http://sonomamg.ucanr.edu/History/Victory_Gardens_in_World_War_II/.

Reclaiming Seeds and Knowledge in Indian Country

- Black, Sandra A. "Diabetes, Diversity, and Disparity: What Do We Do with the Evidence?" *American Journal of Public Health* 92, no. 4 (April 2002). https://dx.doi.org/10.2105%2Fajph.92.4.543.

- Cook, Noble David. *Born to Die; Disease and New World Conquest 1492-1650.* Cambridge: Cambridge University Press, 1998.

- 47. Fox, Mary Kay, William Hamilton, and Biing-Hwan Lin. *Effects of Food Assistance and Nutrition Programs on Nutrition and Health: Volume 3, Literature Review.* Washington, U.S. Department of Agriculture Economic Research Service, 2004. https://www.ers.usda.gov/publications/pub-details/?pubid=46574.

- Inter Tribal Council of Arizona. "Havasupai Tribe." Accessed April 26, 2020. https://itcaonline.com/member-tribes/havasupai-tribe/.

- Kappeler, Victor E., PhD. Police Studies Online at Eastern Kentucky University. "A Brief History of Slavery and the Origins of American Policing." Accessed April 26, 2020. https://plsonline.eku.edu/insidelook/brief-history-slavery-and-origins-american-policing.

- Opala, Joseph A. "The Gullah: Rice, Slavery, and the Sierra Leone-American Connection." Gilder Lehrman Center for the Study of Slavery, Resistance, and Abolition at Yale University. Accessed April 26, 2020. https://glc.yale.edu/gullah-rice-slavery-and-sierra-leone-american-connection.

- Public Broadcasting Service. "Indian Removal." Accessed April 26, 2020. https://www.pbs.org/wgbh/aia/part4/4p2959.html.

- Spanakis, Elias K. and Sherita Hill Golden. "Race/Ethnic Difference in Diabetes and Diabetic Complications." *Current Diabetes Reports* 13, no. 6 (2013). https://dx.doi.org/10.1007%2Fs11892-013-0421-9.

- Wendell Cox Consultancy. "Greater London, Inner London & Outer London Population & Density History." Accessed April 26, 2020. http://www.demographia.com/dm-lon31.htm.

A Seat at the (Farm-to-) Table: Building an Inclusive Future and Claiming Our Power

- Mount, Jeffrey and Ellen Hanak. *Just the Facts: Water Use in California*. San Francisco: Public Policy Institute of California, 2019. https://www.ppic.org/publication/water-use-in-california/

- Pimental, David. "Soil Erosion: A Food and Environmental Threat." *Environment, Development, and Sustainability* 8, no. 1 (February 2006). https://doi.org/10.1007/s10668-005-1262-8.

- U.S. Congress, House, *Recognizing the duty of the Federal Government to create a Green New Deal* HR 109, 116th Cong., 1st sess., introduced in House February 7, 2019.